AF564713

Wind Electrical System

Wind Electrical System

Dr. D. Prasad

RANDOM PUBLICATIONS
NEW DELHI (INDIA)

Wind Electrical System

ISBN 978-93-5111-823-7

Published in 2016 in India by

RANDOM PUBLICATIONS

4376-A/4B, Gali Murari Lal, Ansari Road
New Delhi-110 002
Phone : +9111-43580356, 011-23289044, 011-43142548
e-mail: sales@randompublications.com,
info@randompublications.com, randomexports@gmail.com

Reprinted 2026

Type Setting by : Friends Media, Delhi-110089

Preface

A wind electric system is made up of a wind turbine mounted on a tower to provide better access to stronger winds. In addition to the turbine and tower, small wind electric systems also require balance-of-system components.

The swept area of a wind turbine is the second most important factor (after the wind resource itself) that determines energy production. The circle "swept" by the blades is the collector area. It's not possible to get a large amount of energy out of a small collector area. Betz' theorem says we can only get about 60% of the energy out of the wind before we start slowing it down too much and actually decreasing performance. In the real world, well-designed machines can achieve about half of that.

Turbines can be divided by orientation, directionality, generating mode, and by other characteristics. Horizontal-axis wind turbines (HAWTs) are the most common and effective orientation. Vertical-axis wind turbines (VAWTs) may appeal to the uninitiated, but continue to disappoint as far as performance and longevity—both of the machines and the companies. Upwind (the wind hits the turbine before it hits the tower) and downwind (the wind hits the tower before it hits the turbine) designs can both be very effective.

A wind generator tower is very often more expensive than the turbine. The tower puts the turbine up in the "fuel"—the smooth strong winds that give the most energy. Wind turbines should be sited at least 30 feet (9 m) higher than anything within 500 feet (152 m).

Three common types of towers are tilt-up, fixed-guyed, and freestanding. Towers must be specifically engineered for the lateral thrust and weight of the turbine, and should be adequately grounded to protect your equipment against lightning damage. See my article "Wind Generator Tower Basics" in HP105 for information about choosing a tower.

The book assumes no prior knowledge in the field and therefore would also be suitable for readers with a non-electrical-engineering background.

– Author

Contents

1

Wind Power System

INTRODUCTION

Wind is a form of solar energy. Winds are caused by the uneven heating of the atmosphere by the sun, the irregularities of the earth's surface, and rotation of the earth. Wind flow patterns are modified by the earth's terrain, bodies of water, and vegetative cover. This wind flow, or motion energy, when "harvested" by modern wind turbines, can be used to generate electricity.

HOW WIND POWER IS GENERATED

The terms "wind energy" or "wind power" describe the process by which the wind is used to generate mechanical power or electricity. Wind turbines convert the kinetic energy in the wind into mechanical power. This mechanical power can be used for specific tasks (such as grinding grain or pumping water) or a generator can convert this mechanical power into electricity to power homes, businesses, schools, and the like.

WIND POWER

The market for wind turbines is growing by 20 per cent a year, with no sign of slowing up. The secret is not the technology or the price - it's the policy. At 7-12 cents US/kWh, wind energy is competitive with most other forms of energy generation. Globally, by 2007, 94 GW of wind capacity had been installed. By the end of 2011 this had increased by 253 per cent to 238 GW.

GLOBAL HIGHLIGHTS

- In 2000, the World Energy Assessment team estimated that 27 per cent of the world's land-surface has winds that blow at Class 3 or higher, of which 4 per cent can be used for wind farms, for a total of 19,000 TWh.
- Grubb and Meyer estimated the global wind potential to be 50,000 TWh. Both of these studies look only at land-based wind. Offshore potential may increase this to 100,000 TWh, seven times more electricity than the world is currently using.

- In 2005, two Stanford researchers (Archer and Jacobson) found that the global wind potential was many times larger than previously thought, able to produce 200,000 TWh of electricity, more than the total primary energy used in the world in 2006 (c. 142,000 TWh).
- In 2012, a team of researchers at Stanford University found that wind power could generate up to 1,800 terrawatts of power, which is 100 times the current global power requirement of 18 terrawatts.
- In Germany, 2086 MW of new turbines were installed in 2011 for a total capacity of 29,000 MW, producing 12 per cent of Germany's electricity, thanks to the renewable energy feed-in tariff which gives guaranteed access to the grid and a stable 20 year price favourable to investors.
- In Europe, the southern North Sea has enough wind energy potential to provide three times more electricity than the combined demand of the five countries which have rights to the sea: Britain, Belgium, Holland, Germany and Denmark.
- By the end of 2011, the top countries for electricity generating capacity were China 62 GW, USA 47 GW, Germany 29 GW, and Spain 21 GW.
- Canada increased its wind generating capacity during 2011 to 5,265 MW, placing 9th in the world. Ontario leads with 1,969 MW. BC has 247.5 MW . There are expectations thata further 1500 MW will be installed during 2012.
- In the US, the lower 48 states have sufficient wind potential to produce 10,871 TWh of electricity, three times more than was generated by all means in the entire US in 1998. North Dakota, Texas, and Kansas could provide almost 100 per cent of US electricity needs. Another study suggested that the wind corridors in the Kansas and Nebraska wheat-lands could provide 200 per cent of America's electricity needs.
- With all wind projects, capacity factor measures the percentage of time that a turbine is operating. Older land-based turbines operate at 25 per cent capacity; new ones reach 30 per cent. Offshore wind can reach 40 per cent. At 33 per cent capacity factor, 300 MW of wind energy is the equivalent of 100 MW of hydro operating at 100 per cent capacity factor.

WHAT'S HAPPENING IN BC?

BC has 247.5 MW of installed wind power, following successful bids under BC Hydro's Clean Power Calls:

- Bear Mountain, near Dawson Creek: 102 MW (34 Enercon MW turbines)

- Dokie Wind Project: 144 MW (48 Vestas V-90 turbines)
- Eye of the Wind, Grouse Mountain: 1.5 MW (Leitwind turbine)
- A 99 MW wind farm has been approved and is under construction at Knob Hill on the northern tip of Vancouver Island, with commercial operation to begin in 2012.

In 2012, a new wind energy assessment by Garrad Hassan found a falling cost for wind energy in BC, and that wind energy is the most cost-effective renewable energy technology for large amounts of new generation. The cost of new turbines has fallen, and new wind turbines optimised for sites with lower average wind speeds have resulted in a 34 per cent increase in wind electricity generation in these areas.

According to BC Hydro, 19 of the top 20 most cost-effective wind energy sites in BC are in the Peace Region. The new assessment indicates that wind energy resources outside of the Peace region appear to be more competitive.

WHAT DOES WIND ENERGY COST?

Globally, depending on the site, wind energy averages a levelised cost of $70 to $120 MWh (7 to 12 cents/kWh). In British Columbia, because of mountainous and often remote conditions, the levelised cost for most sites is $90 to $150 per MWh. Canada's Wind Power Production Incentive of 1 cent/ kWh was ended in 2011.

ENVIRONMENTAL MATTERS

Provided there is good public consultation with regard to siting, concerns about noise and visual pollution have generally proven unwarranted. If siting is poorly planned due to inadequate consultation, noise can become a problem, and local objections will arise. Surveys in Britain show a consistent 70 per cent-80 per cent level of support for wind energy, rising up to 94 per cent among people who live near them. A survey of tourists to areas where wind farms are visible found that 91 per cent said the turbines made no difference to their recreational experience.

Concerns about bird deaths have largely disappeared with the larger turbines, slower blade speeds, and proper site consultations with ornithological societies. The US Audubon Society has stated its strong support for wind farms. Studies show that bird fatalities are caused as follows:

- Wind turbines: 1 per 10,000
- Pesticides: 710 per 10,000
- Cats: 1060 per 10,000
- Buildings/Windows: 5020 per 10,000

The real concern for birds is that up to a quarter of all birds could become extinct by 2054 due to global climate change, for which wind energy is one of the solutions. (Nature, 2004)

Danish offshore wind projects researched by biologists revealed a large increase in fishing yields, attributable to the fact that the turbine foundations act as artificial reefs, attracting marine flora and fauna.

TECHNICAL MATTERS

In BC, BC Hydro has such large hydro resources that it is relatively easy to firm up wind power's intermittency by storing surplus power behind the dams when wind is producing more than is needed (eg at night). The development of pumped storage, flow-batteries and other technologies could help with wind power storage. Improved wind forecasting can also improve the switch from one type of generation to another.

SOCIAL, ECONOMIC AND POLITICAL MATTERS

The three great advantages of wind energy are its low price, price stability, and the fact that it is a clean, renewable source of energy that produces no carbon dioxide.

If wind turbines were manufactured in BC, their manufacturing and installation would create 6 jobs per MW, so each 100 MW project would produce 600 full-time jobs. In the USA, wind energy creates 27 per cent more jobs per kilowatt-hour than coal-fired power, and 66 per cent more than natural gas-fired power. A 100 MW installation will typically generate around $850,000 in local purchases. The BC government made a commitment in 2007 that 90 per cent of BC's power would come from clean, renewable sources, and that BC would be self-sufficient in power by 2016.

WIND TURBINES

Wind turbines, like aircraft propeller blades, turn in the moving air and power an electric generator that supplies an electric current. Simply stated, a wind turbine is the opposite of a fan. Instead of using electricity to make wind, like a fan, wind turbines use wind to make electricity. The wind turns the blades, which spin a shaft, which connects to a generator and makes electricity.

Wind Turbine Types

Modern wind turbines fall into two basic groups; the horizontal-axis variety, like the traditional farm windmills used for pumping water, and the vertical-axis design, like the eggbeater-style Darrieus model, named after its French inventor. Most large modern wind turbines are horizontal-axis turbines.

Turbine Components

Horizontal turbine components include:

- Blade or rotor, which converts the energy in the wind to rotational shaft energy;

- A drive train, usually including a gearbox and a generator;
- A tower that supports the rotor and drive train; and
- Other equipment, including controls, electrical cables, ground support equipment, and interconnection equipment.

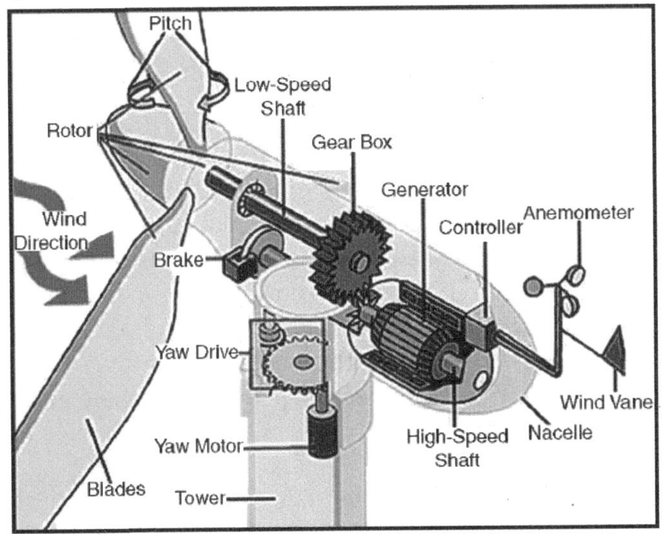

Turbine Configurations

Wind turbines are often grouped together into a single wind power plant, also known as a wind farm, and generate bulk electrical power. Electricity from these turbines is fed into a utility grid and distributed to customers, just as with conventional power plants.

Wind Turbine Size and Power Ratings

Wind turbines are available in a variety of sizes, and therefore power ratings. The largest machine has blades that span more than the length of a football field, stands 20 building stories high, and produces enough electricity to power 1,400 homes.

A small home-sized wind machine has rotors between 8 and 25 feet in diameter and stands upwards of 30 feet and can supply the power needs of an all-electric home or small business. Utility-scale turbines range in size from 50 to 750 kilowatts. Single small turbines, below 50 kilowatts, are used for homes, telecommunications dishes, or water pumping.

WIND ENERGY RESOURCES

Wind energy is very abundant in many parts of the United States. Wind resources are characterised by wind-power density classes, ranging from class 1 (the lowest) to class 7 (the highest). Good wind resources (*e.g.*, class 3 and above, which have an average annual wind speed of at least 13 miles per hour) are found in many locations. Wind speed is a critical feature of wind resources, because the energy in wind is proportional to thecube of the wind speed. In other words, a stronger wind means a lot more power.

ADVANTAGES AND DISADVANTAGES OF WIND-GENERATED ELECTRICITY

A Renewable Non-Polluting Resource

Wind energy is a free, renewable resource, so no matter how much is used today, there will still be the same supply in the future. Wind energy is also a source of clean, non-polluting, electricity. Unlike conventional power plants, wind plants emit no air pollutants or greenhouse gases. According to the U.S. Department of Energy, in 1990, California's wind power plants offset the emission of more than 2.5 billion pounds of carbon dioxide, and 15 million pounds of other pollutants that would have otherwise been produced. It would take a forest of 90 million to 175 million trees to provide the same air quality.

Cost Issues

Even though the cost of wind power has decreased dramatically in the past 10 years, the technology requires a higher initial investment than fossil-fuelled generators. Roughly 80 per cent of the cost is the machinery, with the balance being site preparation and installation. If wind generating systems are compared with fossil-fuelled systems on a "life-cycle" cost basis (counting fuel and operating expenses for the life of the generator), however, wind costs are much more competitive with other generating technologies because there is no fuel to purchase and minimal operating expenses.

Environmental Concerns

Although wind power plants have relatively little impact on the environment compared to fossil fuel power plants, there is some concern over the noiseproduced by the rotor blades, aesthetic (visual) impacts, and birds and bats having been killed (avian/bat mortality) by flying into the rotors. Most of these problems have been resolved or greatly reduced through technological development or by properly siting wind plants.

Supply and Transport Issues

The major challenge to using wind as a source of power is that it is intermittent and does not always blow when electricity is needed. Wind cannot be stored (although wind-generated electricity can be stored, if batteries are used), and not all winds can be harnessed to meet the timing of electricity demands. Further, good wind sites are often located in remote locations far from areas of electric power demand (such as cities). Finally, wind resource development may compete with other uses for the land, and those alternative uses may be more highly valued than electricity generation. However, wind turbines can be located on land that is also used for grazing or even farming.

WIND POWER IN PUMPING WATER

The wind has been used for pumping water for many centuries; it was in fact the primary method used for dewatering large areas of the Netherlands from the 13th century onwards. Smaller windpumps, generally made from wood, for use to dewater polders, (in Holland) and for pumping sea water in salt workings, (France, Spain and Portugal), were also widely used in Europe and are still used in places like Cape Verde.

However the main type of windpump that has been used is the so-called American farm windpump. This normally has a steel, multibladed, fan-like rotor, which drives a reciprocating pump linkage usually via reduction gearing that connects directly with a piston pump located in a borehole directly below. The American farm windpump evolved during the period between 1860 and 1900 when many millions of cattle were being introduced on the North American Great Plains. Limited surface water created a vast demand for water lifting machinery, so windpumps rapidly became the main general purpose power source for this purpose. The US agricultural industry spawned a multitude of windpump manufacturers and there were serious R&D programmes, some sponsored by the US government to evolve better windpumps for irrigation as well as for water supply duties.

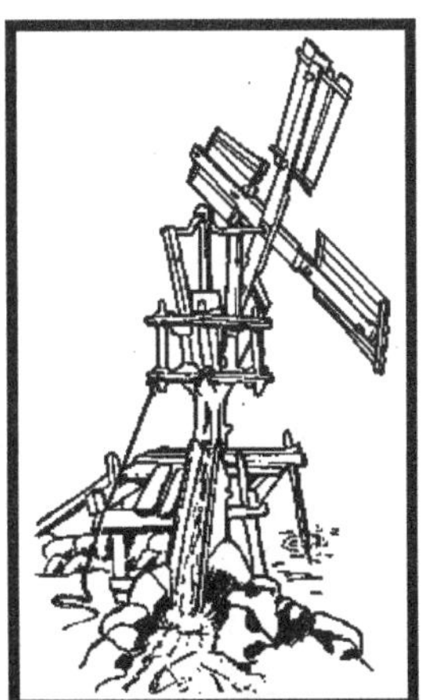

Fig. Wooden indigenous windmill pump for pumping sea water into salt pans on the Island of Sal, Cape Verde

Other "new frontiers" such as Australia and Argentina took up the farm windpump, and to this day an estimated one million steel farm windpumps are in regular use, the largest numbers being in Australia and Argentina. It should be noted that the so-called American Farm Windpump is rarely used today for irrigation; most are used for the purpose they were originally developed for, namely watering livestock and, to a lesser extent, for farm or community water supplies. They tend therefore to be applied at quite high heads by irrigation standards; typically in the 10 to 100m range on boreholes. Large windpumps are even in regular use on boreholes of over 200m depth. Wind pumps have also been used in SE Asia and China for longer than in Europe, mainly for irrigation or for pumping sea water into drying pans for sea salt production.

Fig. All-steel 'American' farm wind pump

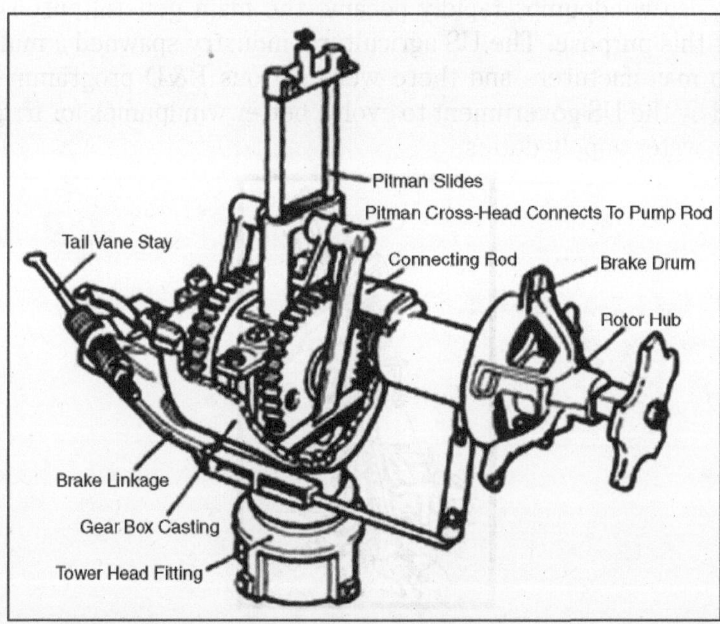

Fig. Gearbox from a typical back-geared 'American' farm windmill

The Chinese sail windpump was first used over a thousand years ago and tens if not hundreds of thousands, are still in use in Hubei, Henan and North Jiangsu provinces. The traditional Chinese designs are constructed from wire-braced bamboo poles carrying fabric sails; usually either a paddle pump or a dragon-spine (ladder pump) is used, typically at pumping heads of less than lm.

Many Chinese windmills rely on the wind generally blowing in the same direction, because their rotors are of fixed orientation. Many hundreds of a similar design of windpump to the Chinese ones are also used on saltpans in Thailand.

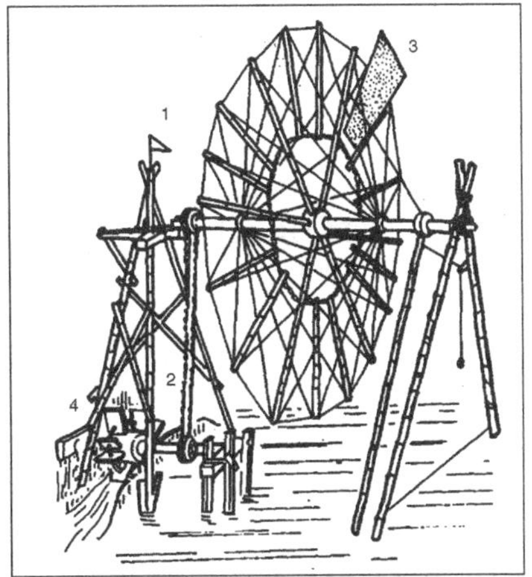

Fig. Chinese chain windmill

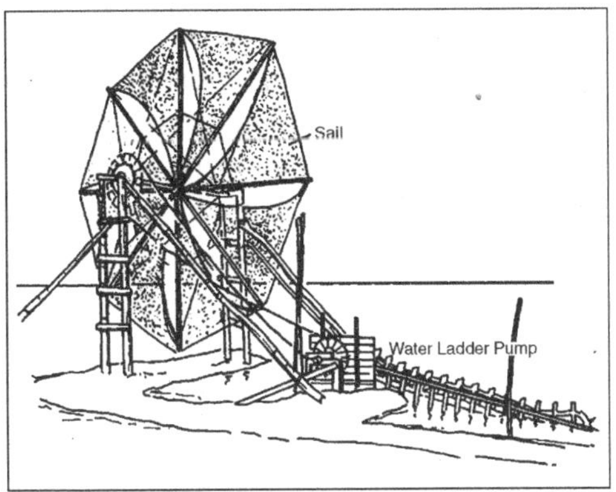

Fig. Thai windpump

Fig. 'Cretan' type of windmill used on an irrigation project in Southern Ethiopia

Some 50 000 windpumps were used around the Mediterranean Sea 40 years ago for irrigation purposes. These were improvised direct-drive variations of the metal American farm windpump, but often using triangular cloth sails rather than metal blades. These sail windmills have a type of rotor which has been used for many centuries in the Mediterranean region, but today is often known as "Cretan Windmills".

During the last 30 years or so, increased prosperity combined with cheaper engines and fuels has generally led farmers in this region to abandon windmills and use small engines (or mains electricity where available). However Crete is well known as a country where until recently about 6 000 windpumps were still in use, mostly with the cloth sailed rig. The numbers of windpumps in use in Crete are rapidly declining and by 1986 were believed to be barely one thousand.

Another branch of wind energy technology began to develop in the late 1920s and early 1930s, namely, the wind-generator or aero-generator. Many thousands of small wind generators, such as the Australian Dunlite, were brought into use for charging batteries which could be used for lighting, and especially for radio communication, in remote rural areas. Such machines can also provide an alternative to a photovoltaic array for irrigation pumping in suitably windy areas, although they have not so far been applied for this purpose in any numbers. Large wind turbines for electricity generation have been (and are being) constructed, the largest being a 5MW (5 000kW) machine under development in West Germany. However, more modest, but still quite large medium sized machines are being installed in large numbers for feeding the local grid notably in the state of California (where over 10 000 medium sized wind generators have been installed in little more than 3 years for feeding the grid) and in Denmark. A typical modern 55kW, 15m diameter Windmatic wind turbine, from Denmark. Machines of this size may in future be of considerable relevance for larger scale irrigation pumping than is feasible with more traditional mechanical windpumps.

STATE-OF-THE-ART

There are two distinct end-uses for windpumps, namely either irrigation or water supply, and these give rise to two distinct categories of windpump because the technical, operational and economic requirements are generally different for these end uses. That is not to say that a water supply windpump cannot be used for irrigation (they quite often are) but irrigation designs are generally unsuitable for water supply duties.

Most water supply windpumps must be ultra-reliable, to run unattended for most of the time (so they need automatic devices to prevent overspeeding in storms), and they also need the minimum of maintenance and attention and to be capable of pumping water generally from depths of 10m or more. A typical farm windpump should run for over 20 years with maintenance only once every year, and without any major replacements; this is a very demanding technical requirement since typically such a wind pump must average over 80 000 operating hours before anything significant wears out; this is four to ten times the operating life of most small diesel engines or about 20 times the life of a small engine pump. Windpumps to this standard therefore are usually industrially manufactured from steel components and drive piston pumps via reciprocating pump rods. Inevitably they are quite expensive in relation to their power output, because of the robust nature of their construction. But American, Australian and Argentinian ranchers have found the price worth paying for windpumps that achieve high reliability and minimum need for human intervention, as this is their main advantage over practically any other form of pumping systems.

Fig. 2kW Dunlite wind electricity generator

Fig. 55kW Windamatic wind electricity generator

Irrigation duties on the other hand are seasonal (so the windmill may only be useful for a limited fraction of the year), they involve pumping much larger volumes of water through a low head, and the intrinsic value of the water is low. Therefore any windpump developed for irrigation has to be low in cost and this requirement tends to overide most other considerations. Since irrigation generally involves the farmer and/or other workers being present, it is not so critical to have a machine capable of running unattended. Therefore windmills used for irrigation in the past tend to be indigenous designs that are often improvized or built by the farmer as a method of low-cost mechanisation. If standard farm windpumps are used for irrigation, usually at much lower heads than are normal for water supply duties, there are quite often difficulties in providing a piston pump of sufficient diameter to give an adequate swept volume to absorb the power from the windmill. Also most farm windpumps have to, be located directly over the pump, on reinforced concrete foundations, which usually limits these machines to pumping from wells or boreholes rather than from open water.

A suction pump can be used on farm windmills with suction heads of up to about 5-6m from surface water. Most indigenous irrigation windpumps, on the other hand, such as those in China, use rotary pumps of one kind or another which are more suitable for low heads; they also do not experience such high mechanical forces as an industrial windpump. Attempts have been made recently to develop lower cost steel windpumps that incorporate the virtues of the heavier older designs. Most farm-windpumps, even though still in commercial production, date back to the 1920s or earlier and are therefore unecessarily heavy and expensive to manufacture, and difficult to install properly in remote areas. Recently various efforts have been made to revise the traditional farm windpump concept into a lighter and simpler modern form.

The "IT Windpump", which is half the weight of most traditional farm windpump designs of a similar size, and is manufactured in Kenya as the "Kijito" and in Pakistan as the "Tawana". The latter costs only about half as much as American or Australian machines of similar capability. It is possible therefore that through developments of this kind, costs might be kept low enough to allow the marketing of all steel windpumps that are both durable like the traditional designs, yet cheap enough to be economic for irrigation.

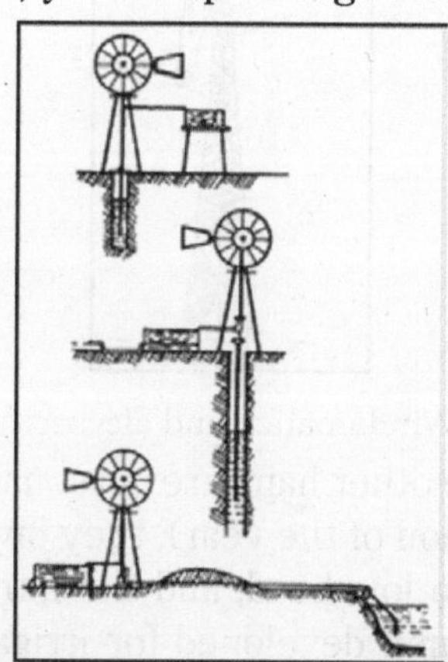

Fig. Typical farm windpump installation configurations

A. borehole to raised storage tank

B. well to surface storage tank
C. surface suction pump

Fig. IT windpump, made in Kenya as the 'Kijito' and in Pakistan as the 'Tawana'

CONVERTING WIND POWER TO SHAFT POWER

There are two main mechanisms for converting the kinetic energy of the wind into mechanical work; both depend on slowing the wind and thereby extracting kinetic energy. The crudest, and least efficient technique is to use drag; drag is developed simply by obstructing the wind and creating turbulence and the drag force acts in the same direction as the wind.

Some of the earliest and crudest types of wind machine, known generically as "panamones", depend on exposing a flat area on one side of a rotor to the wind while shielding (or reefing the sails) on the other side; the resulting differential drag force turns the rotor. The other method, used for all the more efficient types of windmill, is to produce lift. Lift is produced when a sail or a flat surface is mounted at a small angle to the wind; this slightly deflects the wind and produces a large force perpendicular to the direction of the wind with a much smaller drag force.

It is this principle by which a sailing ship can tack at speeds greater than the wind. Lift mainly deflects the wind and extracts kinetic energy with little turbulence, so it is therefore a more efficient method of extracting energy from the wind than drag.

It should be noted that the theoretical maximum fraction of the kinetic energy in the wind that could be utilised by a "perfect" wind turbine is approximately 60 per cent. This is because it is impossible to stop the wind completely, which limits the percentage of kinetic energy that can be extracted.

HORIZONTAL AND VERTICAL AXIS ROTORS

Windmills rotate about either a vertical or a horizontal axis. All the windmills illustrated so far, and most in practical use today, are horizontal axis, but research is in progress to develop vertical axis machines. These have the advantage that they do not need to be orientated to face the wind, since they present the same cross section to the wind from any direction; however this is also a disadvantage as under storm conditions you cannot turn a vertical axis rotor away from the wind to reduce the wind loadings on it.

There are three main types of vertical axis windmill. Panamone differential drag devices (mentioned earlier), the Savonius rotor or "S" rotor and the Darrieus wind turbine. The Savonius rotor consists of two or sometimes three curved interlocking plates grouped around a central shaft between two end caps; it works by a mixture of differential drag and lift.

The Savonius rotor has been promoted as a device that can be readily improvized on a self-build basis, but its apparent simplicity is more perceived than real as there are serious problems in mounting the inevitably heavy rotor securely in bearings and in coupling its vertical drive shaft to a positive displacement pump (it turns too slowly to be useful for a centrifugal pump).

However the main disadvantages of the Savonius rotor are two-fold:

1. It is inefficient, and involves a lot of construction material relative to its size, so it is less cost-effective as a rotor than most other types;
2. It is difficult to protect it from over-speeding in a storm and flying to pieces.

The Darrieus wind turbine has airfoil cross-section blades (streamlined lifting surfaces like the wings of an aircraft). These could be straight, giving the machine an "H"-shaped profile, but in practice most machines have the curved "egg-beater" or troposkien profile as illustrated.

The main reason for this shape is because the centrifugal force caused by rotation would tend to bend straight blades, but the skipping rope or troposkien shape taken up by the curved blades can resist the bending forces effectively. Darrieus-type vertical axis turbines are quite efficient, since they depend purely on lift forces produced as the blades cross the wind (they travel at 3 to 5 times the speed of the wind, so that the wind meets the blade at a shallow enough angle to produce lift rather than drag).

The Darrieus was predated by a much cruder vertical axis windmill with Bermuda (triangular) rig sails from the Turks and Caicos Islands of the West Indies. This helps to show the principle by which the Darrieus works, because it is easy to imagine the sails of a Bermuda rig producing a propelling force as they cut across the wind in the same way as a sailing yacht; the Darrieus works on exactly the same principle.

Fig. Savonius Rotor vertical-axis windpump in Ethiopia.

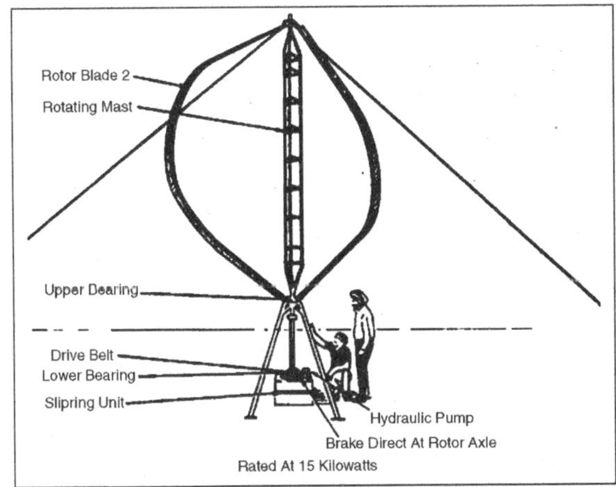

Fig. Typical Troposkien shaped Darrieus vertical axis wind turbine

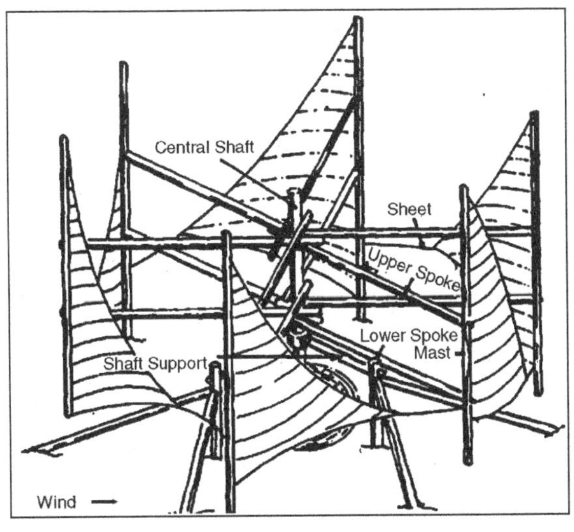

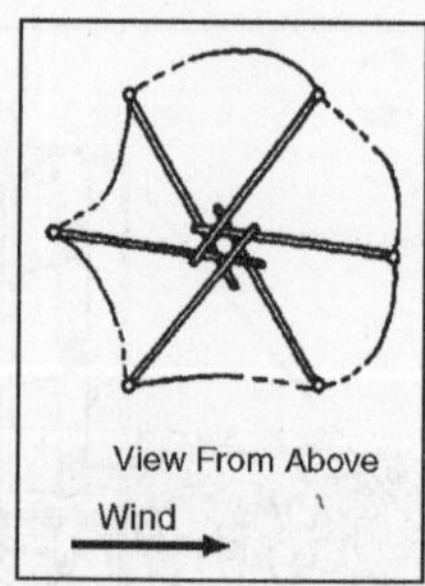

Fig. Turks and Caicos islands vertical-axis sail rotor

There are also two main types of Darrieus wind turbine which have straight blades; both control overspeed and consequent damage to the blades by incorporating a mechanism which reefs the blades at high speeds. These are the Variable Geometry Vertical Axis Wind Turbine (VGVAWT) developed by Musgrove in the UK and the Gyromill Variable Pitch Vertical Axis Wind Turbine (VPVAWT), developed by Pinson in the USA. Although the Musgrove VGVAWT has been tried as a windpump by P I Engineering, all the current development effort is being channelled into developing medium to large electricity grid-feeding, vertical-axis wind-generators, of little relevance to irrigation pumping.

Vertical axis windmills are rarely applied for practical purposes, although they are a popular subject for research. The main justification given for developing them is that they have some prospect of being simpler than horizontal axis windmills and therefore they may become more cost-effective. This still remains to be proved.

Most horizontal axis rotors work by lift forces generated when "propeller" or airscrew like blades are set at such an angle that at their optimum speed of rotation they make a small angle with the wind and generate lift forces in a tangential direction. Because the rotor tips travel faster than the roots, they "feel" the wind at a shallower angle and therefore an efficient horizontal axis rotor requires the blades to be twisted so that the angle with which they meet the wind is constant from root to tip. The blades or sails of slow speed machines can be quite crude but for higher speed machines they must be accurately shaped airfoils; but in all three examples illustrated, the principle of operation is identical.

2

Wind Energy Technology and Wave Motion

WIND ENERGY

There is little potential for wind energy in Peru. Although no map exists that illustrates wind patterns in the Peruvian forest, recent reconnaissance of the High Selva in San Martin, Pucallpa, and Satipo did not detect winds with energy-producing potential. Nevertheless, before discarding this option, and taking into account its unpredictability, wind velocity should be evaluated where wind is being considered a potential energy-producer. Wind's application for mechanical (mills) or electrical (aerogenerators) purposes would depend on the presence of continual wind; the demand (water to be pumped or KW required); the design and dimensions of the equipment and; whether equipment is produced nationally or locally.

CONFLICTS AND INTERACTIONS BETWEEN ALTERNATIVE ENERGY USE AND OTHER SECTORS

Summary of Non-conventional Energy Alternatives

Technology	*Process*	*Raw Material*	*Product*	*By-Product*	*State of the Art*	*Applications*	*Applicability in the Central Selva*
1. Hydroenergy	P.G.H.	Water Courses and Waterfalls	Electricity	-	Commercial	Rural Electrification	The majority of its present and future populations
	Waterwheels	Water Courses and Waterfalls	Mechanical Energy	-	Commercial	Cottage and small industry	Sawmills, carpentry shops, grain mills, sugar mills, etc.
	Hydraulic Rams	Water Courses	Mechanical Energy		Commercial	Pumping of water for domestic and other purposes	Homes and isolated lodging establishments on slopes near rivers
2. Biomass	Direct Combustion	Wood and wood residues	Heat, steam mechanical	Smoke, ash	Commercial	Domestic, rural and industrial	Cooking food, dehydrating agricultural products, ceramic and brick-making ovens, industrial production of paper, operating sawmills, etc.
	Thermo-conversion	Wood, cellulos residues	Charcoal, metallurgical coke	(Phenols) Tar. Methanol acetic acid	Commercial	Domestic, rural metallurgical, industrial	Id., also in steel-making and generating electricity
			Wood gas	Ash. CO_2	Commercial and experimental	Rural and Industrial	Ovens, boilers, and industrial engines, generating electricity
			Methanol	Ash, CO_2	Experimental	Industry and Transport	Chemical industry, vehicles

	Alcohol Fermentation	Sugar cane, manioc, wood, etc.	Ethanol lignin	CO_2, pulp, wine, fusel oil neutralized acid acid	Commercial and experimental (wood ethanol)	Transport, metallurgy, and industry	Gasoline-powered vehicles, foundries, chemical industry
Biomass	Anaerobic Fermentation	Organic, animal and plant waste	Biogas (methane)	Fertilizer, Environmental Sanitation	Commercial and small scale	Energy for domestic, rural and industrial (experimental) use	Cooking food, heating, lighting refrigeration, internal combustion engines, turbine/operation
Solar	Low-level Thermal	Solar Radiation	Heat applied to air and water	Reduction of accessible land	Commercial and experimental	- Dehydrating agricultural products - Heat for chicks	Drying rice, etc.
	High-level thermal production	Solar Radiation	Concentrated heat that generates steam and electricity	Reduction of accessible land	Experimental	Pumps, industrial ovens, electricity	None for the short and medium term
	Photovoltaic	Solar Radiation	Continous electrical current	.ID	Experimental, nearly commercial	- Domestic - Pumping - Telecommunications in remote localities	Wide applicability in colonies, if affordable equipment is available
Wind	Wind-driven	Wind	Mechanical energy	-	Commercial	Water pumping Grain mills, etc.	Little, because of scarcity of wind
	Aero-generators	Wind	Continuous electricity		Commercial (low power) and experimental (high power)	Continuous electricity for domestic use	Little, because of scarcity of wind

The livestock sector can only benefit from biogas techniques which, among other things, use animal wastes. Because Central Selva soils are poor in phosphorus, phosphate needs to be imported and applied for good grass growth. One way to provide phosphates is to use the livestock manure that now is dumped into water courses and lost.

Biogas can replace kerosene and propane gas in refrigeration and in providing heat for chicken and swine breeding operations (the largest farm in the Central Selva is presently using this technology). It can provide heat and electricity to the human settlements associated with livestock operations. Biofertilizer can also be partially recycled in the animals' diets through the use of digesters.

Animal manure used in aquaculture should be pre-treated aerobically, as this improves its quality as food and reduces contamination risks. This technique is widely utilised in China in carp and tilapia pisciculture.

CONFLICTS AND INTERACTIONS WITH THE USE OF FOREST RESOURCES

Forest operations actually complement biomass energy production. Both the wood remaining in the countryside and the residues of forest production can be used much more efficiently than they are today.

The Central Selva region accounts for 19 percent (132,000 m^3) of national wood production. In a recent survey it contained 116 sawmills, 22 parquet factories, three wood veneer factories, 40 factories that produced cartons, and one that made paper. Wood residues that could produce biomass energy are found at these industries, but today 50 percent of the wood sent through these sawmills is lost and represents a substantial loss of energy and money. Residues are burned, dumped into rivers, or, as at the Pucallpa factory, burned in special ovens which do not exploit the heat produced. The energy producers and the forestry sector need to find ways to work together to achieve the important goal of efficient energy use.

CONFLICTS WITH AND COMPLEMENTS TO AGRICULTURE

Agricultural residues are excellent for use in anaerobic fermentation, which complements biogas production as it returns the necessary nutrients to the land. In addition, the use of manioc surpluses and the production of hydrocarbon-containing ethanol can help stabilise farmers' prices. Although oleaginous seeds can be used both as food and diesel substitute fuel, food production takes precedence.

CONFLICTS WITH AND COMPLEMENTS TO CONSERVATION

Large scale biogas production requires the introduction of exotic tree species (or monocultures of high-yield species), which can profoundly modify local ecosystems. These plantations can affect, in unpredictable ways, some economically-important pursuits, as well as such native subsistence activities as hunting and fishing. If acid hydrolysis of wood is used to produce ethanol, the acid remaining at the end of the process has to be neutralised, even though it is tempting to use low-yield sugar fields to dispose of these acids without first neutralising them. This, however, can so seriously injure aquatic and plant life at the dumping sites that all life can disappear from the rivers, as actually happened when acid metallurgical wastes were dumped into the Mantaro river.

Pyrolysis operations in which the acids are not recovered can also significantly pollute the atmosphere with escaping tar, methanol, acetic acid, and other vapors. Pyrolysis by-product and effluent disposal needs to be regulated and vapor-condensing units should be used.

Small-scale conflicts are easy to find where biogas operations are not properly managed. If the user does not allow time for anaerobic degradation (because of climatic and operating conditions), the effluent applied as fertilizer can contain pathogenic parasites, especially if the effluent contains human excrement.

THE IMPORTANCE OF WIND SPEED

Kinetic energy in wind can be captured by wind turbines and converted to mechanical energy. Generators produce electricity from the mechanical energy. Simply, wind turbines work like a fan operating backwards. Instead of electricity making the blades turn to blow wind from a fan, wind turns the blades in a turbine to create electricity.

Wind turbines range in size from a few hundred watts to as large as several megawatts. The amount of power produced from a wind turbine depends on the length of the blades (or the term of swept area) and the speed of the wind. The power in the wind is proportional to the cube of the wind speed; the general formula for power in the wind is:

$$P = 1/2 * p * A * V^3$$

where P is the power available in watts, ñ is the density of air (which is approximately 1.2kg/m^3 at sea level), A is the cross-section (or swept area of a windmill rotor) of air flow of interest and V is the instantaneous free-stream wind velocity. Because of this cubic relationship, the power availability is extremely sensitive to wind speed. A doubling the wind speed increases the power availability by a factor of eight. Even a small variation in wind speed converts to a substantial difference in power output. The same turbine on a site with an average wind speed of 8 m/s will produce twice as much as electricity as an on a site with 6 m/s . The usual cut in speed is 5 m/s and full-load attained above 12 m/s, while the usual cut out speed is 25 m/s[1]. Thus developers expend considerable effort to identify and secure the sites which are most consistently in the optimum range.

NREL divides wind speeds into wind power classes designated Class 1 (lowest) through Class 7 (highest). Class 2 and above wind speeds can provide sufficient energy to drive a small wind turbine. Utility sized turbines usually need at least Class 3 wind conditions to operate.

Table. Wind Power Classes at 10 m and 50 m Elevation.

Power class	10 m Wind speed (m/s)	10 m Power Density (W/m^2)	50 m Wind speed (m/s)	50 m Power Density (W/m^2)
1	0-4.4	0-100	1-5.6	0-200
2	4.4-5.1	100-150	5.6-6.4	200-300
3	5.1-5.6	150-200	6.4-7.0	300-400
4	5.6-6.0	200-250	7.0-7.5	400-500
5	6.0-6.4	250-300	7.5-8.0	500-600
6	6.4-7.0	300-400	8.0-8.8	600-800
7	7.0-9.4	400-1,000	8.8-11.9	800-2,000

Wind speed is partly a function of height and generally weaker near the ground due to friction between earth's surface and air flow. So placing turbines on hills and on large towers gives access to higher wind speeds. A taller tower not only makes it possible to reach faster winds but also accommodate a bigger rotor for a larger swept area.

All these factors have driven manufacturers to make ever bigger turbines. The turbines of the mid-1990s swept ten times the area of earlier machines . The size of wind turbines has doubled approximately every 4~5 years . Turbines with an installed generator capacity of 5 to 6 MW and a diameter of 110-120 m diameter are running as prototype.

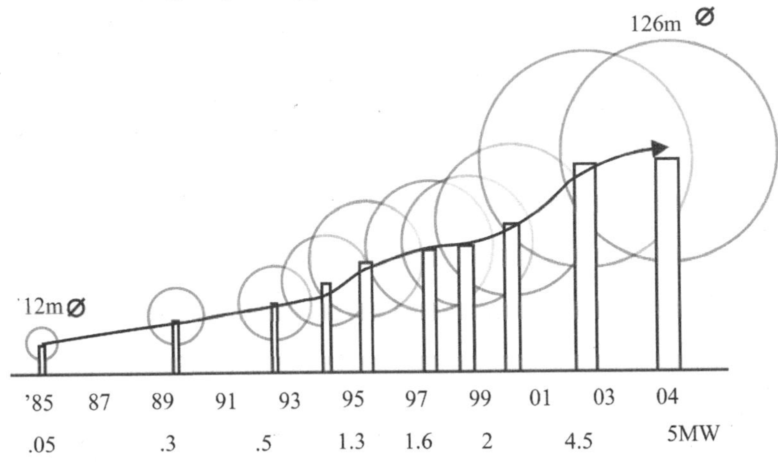

Intermittence

Wind speed variation has system-wide effects for the electricity generation sector. Wind speed can decrease or increase by a factor of two very rapidly.

Each time this happens, generation from a 'wind carpet' – namely, the total number of turbines in a relevant geographical area – decreases or increases by a factor of eight. Fluctuation in wind availability leads to sudden drop-outs and surges in electricity supply, requiring 'up regulation' and 'down regulation' by conventional generating plants . The bigger the 'wind carpet', the more pronounced these effects are.

This creates the problem of 'intermittence', which has two different components. The first is the total absence of wind energy – and therefore of generation – during high pressure events. The second is rapid up or down variation in wind speeds and power output. The first can be predicted by weather forecasts with increasing accuracy. Predictions of the second are improving – due to better methodologies and tools (epically under short time-frames) – but will always remain a problem because wind speed variation is inherently a stochastic phenomenon

Operation

Turbines produce direct current (DC) or alternating (AC) power, depending on the generator. (In our case the prime mover is the rotor.) However, neither way is 100 percent efficient at transferring wind power. The rotor will deliver more power to the generator than the generator produces as electricity. This leads to another fundamental consideration on the size of wind turbines.

The size of a generator indicates only how much power the generator is capable of producing if the wind turbine's rotor is big enough, and if there's enough wind to drive the generator at the right speed. Thus further confront the fact that a wind turbine's size is primarily governed by the size of its rotor.

Wind speeds are crucial for generating utility-compatible electricity. To adapt wind speed variation, electrical generators can be operated either at variable speed or at constant speed. In the first case, the speed of the wind turbine rotor varies with the wind; in the second, the speed of the wind turbine rotor remains relatively constant as wind speed fluctuates.

In nearly all small wind turbines, the speed of the rotor varies with wind speeds. This simplifies the turbines' control while improving aerodynamic performance. When such wind machines drive an alternator, both the voltage and frequency vary with wind speed.

The electricity they produced isn't compatible with the constant-voltage, constant-frequency AC produced by the utility. Electricity from these wind turbines cannot be used in most our daily equipments. The output from these machines must be treated or conditioned first, usually equipped with features to produce correct voltage and constant frequency compatible with the loads.

Although nearly all medium-size wind turbines, such as the thousands of machines installed in California during the early 1980s, operated at constant speed by driving standard, off-the-shelf induction generators, a number of manufacturers of megawatt-size turbines today have switched to variable-speed operation. Many of these use a form of induction generators, which may improve aerodynamic performance.

Estimating output

Annual energy production (AEP[2]) is calculated by applying the predicted wind distributions for a given site to the power performance curve of a particular wind turbine. The site wind distributions are normally based on a Rayleigh distribution that describes how many hours (or probability) each year the wind at a given site blows at a particular wind velocity.

The second step in this process requires the power curve for the chosen turbine. An example of a power curve for a 1.5-MW turbine that is characteristic of current technology.

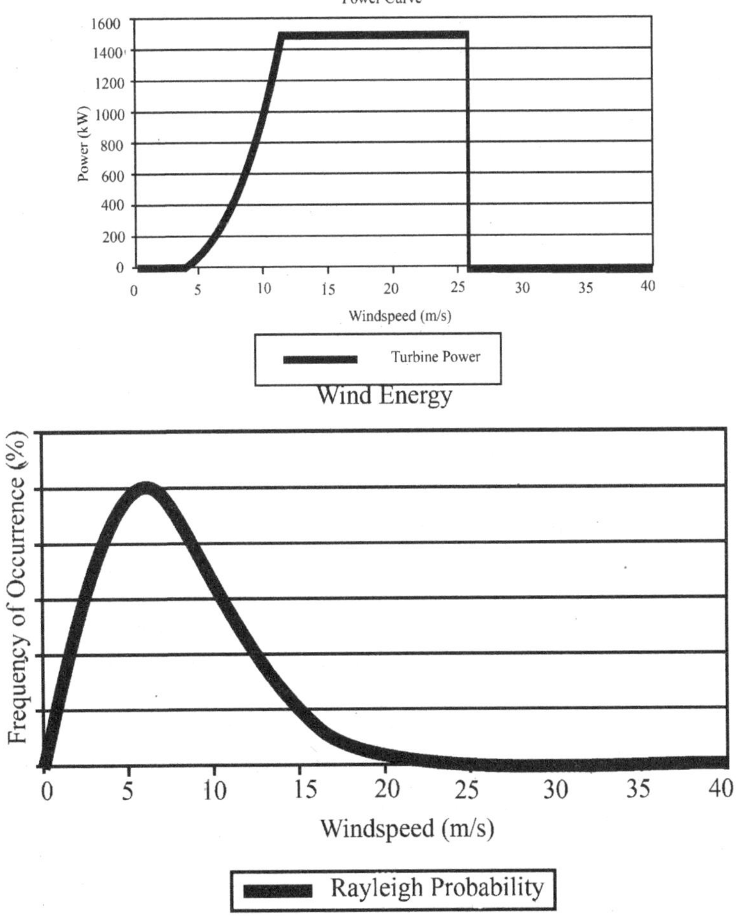

Fig . Power Curve Method of Calculating Annual Energy Output

What may not be stated upfront is that wind speeds of 10-25 m/s are necessary to reach a 1,500 kW output.

The power output of a turbine for wind speeds must be determined specific to a specific site. Wind turbine developers can properly install a turbine that is well-suited for each site.

The product of first two curves will be a curve. By integrating the area under this curve, it is possible to determine the annual energy production. For a more accurate calculation it is necessary to account for both the mechanical and electrical power conversion efficiency, which varies at different turbine power level losses as described in the operating characteristics above, and the projected machine availability.

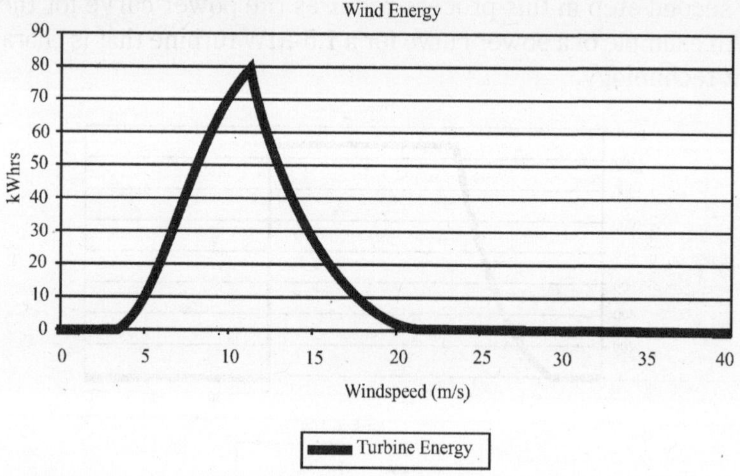

Fig. Power Curve Method of Calculating Annual Energy Output

Conclusion

Wind turbine technology is in good health: the availability of turbines is ~98 per cent, which means that during 2 per cent of the time they cannot produce due to maintenance or failures. In general, elementary design rules dictate that the bigger the turbines are to deploy as much as possible.

In United States, the overall potential is vast. Wind power energy has been estimated as one of the country's most abundant energy resources. About one-fourth of the total land area of United States has winds powerful enough to generate electricity as cheaply as natural gas or coal at today's prices. The wind energy potential in all of the top states – North Dakota, Kansas, South Dakota, Montana, and Nebraska – is principle sufficient to provide all the electricity the country's current uses. In fact, 20 per cent of the land areas in these Midwest and the Rocky Mountain states belong to NREL Class 4, which are sufficient to make your wind energy business profitable. However, few of those potential good sites with a high wind speed have been fully developed yet. In what follows, we will point to how wind energy development is likely to be cost-effective. Albeit the one-time investment on turbines is high, the margin cost per output will be relative small. In a certain future, with growing large turbine machine, supporting technologies and governmental incentive instruments (subsidies and/or tax credits), electricity produced by giant wind turbines in Midwest will be market competitive and become candidate solution to the country's energy independence.

Table. Comparison of Wind Energy Potential vs. Installed Capacity in Selected Contiguous States (GWh/year).

State	Wind Potential	Installed Capacity
North Dakota	1,210	146
Kansas	1,070	227
South Dakota	1,030	97
Montana	1,020	1
Nebraska	868	31
Wyoming	747	643
Oklahoma	725	353

A COST ANALYSIS OF WIND ENERGY

Specifically, we consider the proposed Frontier Line, an HVDC line that would transmit large amounts of electricity to Western cities from remote generation sites. We then analyze the cost of wind-generated electricity under both scenarios. We find that wind-generated electricity is not competitive in nearby energy markets, but is competitive to service Californian demand. Finally, we consider the impact of two polices would have on the economic viability of wind energy: the Renewable Electricity Producer Tax Credit and a cap-and-trade.

THE AVAILABILITY OF WIND RESOURCES

One legitimate drawback to wind-based electricity generation is that the inputs to production are geographically fixed. Unlike fossil fuels or uranium, wind resources cannot be extracted from the earth and transported to demand sites. Instead, they must be converted to electricity at the site where they are found, and the electricity must be transmitted to demand sites.

This limitation is not inherently problematic, because in many scenarios wind resources are very close to demand sites. Denmark, for example, receives a large proportion of their energy from offshore winds that are quite close to demand sites. The U.S. also has abundant offshore wind resources, and with 53 per cent of the US Population living in a coastal county, many electricity demand sites have a large potential supply of wind-generated electricity nearby.

The map below is published by the National Renewable Energy Laboratory (NREL), a facility operated by the U.S. Department of Energy. It illustrates the availability of wind resources at different wind power classes (WPCs). Current technology limits wind-based electricity generation to WPCs of 4 and above, which roughly corresponds to winds average speeds greater than 12.5 mph.

The remoteness of wind resources adds to the cost of wind-generated electricity in two ways: 1) it necessitates large capital investments in transmission lines; and 2) it increases the size of transmission losses. However, the extent of these costs are poorly understood. Our cost analyses will consider these costs, and their effect on the commercial viability of wind energy.

TRANSMISSION TO LOCAL ELECTRICITY MARKETS

Our first scenario estimates the cost of wind-generated electricity delivered to the nearest electricity demand site. Due to the particularly remote location of terrestrial wind resources, we hypothesized that transmission costs would form a significant portion of overall electricity costs, and thus attempted to minimize transmission costs by calculating the shortest distance from each wind resource site to the nearest electricity demand site.

Wind resource site data was made available by NREL in the form of GIS (Geographic Information System) files. These files contain the Wind Power Class data for each square kilometer of land in the Western states. Because NREL's wind resource database is still being built, data files were unavailable for many states of interest, including Minnesota, Iowa, Kansas, Oklahoma, and Texas. However, many of the states with the best wind resources were made available for analysis. These included North and South Dakota, Nebraska, Montana, Wyoming, and Colorado.

Electricity demand sites were selected as the top 25 Metropolitan Statistical Areas, according to the 2007 estimates by the U.S. Census Bureau. The absence of city-level electricity consumption data makes it impossible to verify if population rank matches up with electricity consumption rank. However, we can be confident that each of the 25 most populated MSA consume a large amount of electricity, and are therefore potential markets for remote wind-generated electricity.

Procedurally, we used ArcGIS software to calculate the centroid of each MSA. We then created a map layer of Thiessen polygons, polygons that contain exactly one MSA and all the points nearest to that MSA. All wind resources sites falling within a particular Thiessen polygon would transmit electricity to the MSA within the polygon, because that is the nearest major electricity demand site.

The two graphics below display the results of this analysis. The first show the Thiessen polygon projection for the top 25 MSA in America. It is apparent from this projection that Denver, Colorado is the nearest MSA for the majority of suitable wind sites.

EXPLOITING PRICE VARIATION

Our second scenario predicts that wind-generated electricity will be transmitted to consumption sites where it can exploit variation in prices. Electricity prices vary considerably across states, as is evident from the map on the following page of average 2006 prices:

The average price of electricity is very low in states where wind resources are sited. Thus, although these markets are nearby, wind energy faces stiff competition when competing in local markets (Wyoming, a state with abundant wind resources, had the third lowest average price of any state in 2006, at just 5.27 cents/kWh). However, a long-distance transmission line could carry wind-generated electricity to higher-price markets where it might prove more competitive. At 12.82 cents/kWh, the Californian electricity market looks particularly appealing. Such a project would require transmission beyond the 500 mile break-even point between HVAC and HVDC, leading policymakers to favour the use of an HVDC line.

In fact, proposals to build a long-distance HVDC transmission line are currently being heard. Western governors have formed a task group to research many transmission investments, including the Frontier Line, a high-voltage transmission line that would connect Wyoming energy supply to California loads.

With energy demand increasing at 2 per cent a year, California must add 1,000 MW of electric capacity every year to keep prices constant. Apart from needing additional electricity, California also favours renewable energy. In 2002 the state established the Renewable Portfolio Standard Programme, marking a commitment to achieve 20 per cent of electricity from renewable sources by 2017.

With just 10.9 per cent of electricity coming from renewables, there is significant room for expansion to meet this commitment. California is therefore an eager market for Wyoming wind-generated electricity.

The Frontier Line would address the two major disadvantages of wind energy. For one, an HVDC line minimizes transmission costs, thus addressing the issue of wind resource availability. Secondly, the Frontier Line is well-positioned to weather wind resource variability due to it's origin in Wyoming. Not only does Wyoming have abundant wind resources; it also houses the nation's largest supply of coal. When the wind is not blowing, coal plants can generate electricity to keep the transmitted supply constant at 100 per cent of line capacity.

Thus, wind-based electricity generation can be backed up by coal-based generation, in much the same way that the Danish wind energy system decreases variability by relying on the German coal-fired grid. Therefore, the Frontier Line may avoid the two major pitfalls of wind energy, availability and variability of wind resources.

WIND WAVE

In fluid dynamics, wind waves or, more precisely, wind-generated waves are surface waves that occur on the free surface of oceans, seas, lakes, rivers, and canals or even on small puddles and ponds. They usually result from the wind blowing over a vast enough stretch of fluid surface. Waves in the oceans can travel thousands of miles before reaching land. Wind waves range in size from small ripples to huge waves over 30 m high. When directly generated and affected by local winds, a wind wave system is called a wind sea. After the wind ceases to blow, wind waves are called swells. More generally, a swell consists of wind-generated waves that are not-or are hardly-affected by the local wind at that time. They have been generated elsewhere or some time ago. Wind waves in the ocean are called ocean surface waves.

Wind waves have a certain amount of randomness: subsequent waves differ in height, duration, and shape with limited predictability. They can be described as a stochastic process, in combination with the physics governing their generation, growth, propagation and decay-as well as governing the interdependence between flow quantities such as: the water surface movements, flow velocities and water pressure. The key statistics of wind waves (both seas and swells) in evolving sea states can be predicted with wind wave models. Tsunamis are a specific type of wave not caused by wind but by geological effects. In deep water, tsunamis are not visible because they are small in height and very long in wavelength. They may grow to devastating proportions at the coast due to reduced water depth.

WAVE FORMATION

The great majority of large breakers one observes on a beach result from distant winds. Five factors influence the formation of wind waves:

- Wind speed
- Distance of open water that the wind has blown over (called the fetch)
- Width of area affected by fetch
- Time duration the wind has blown over a given area
- Water depth

All of these factors work together to determine the size of wind waves. The greater each of the variables, the larger the waves. Waves are characterized by:

- Wave height (from trough to crest)
- Wave length (from crest to crest)
- Wave period (time interval between arrival of consecutive crests at a stationary point)
- Wave propagation direction

Waves in a given area typically have a range of heights. For weather reporting and for scientific analysis of wind wave statistics, their characteristic height over a period of time is usually expressed as significant wave height. The significant wave height is also the value a "trained observer" (e.g. from a ship's crew) would estimate from visual observation of a sea state. Given the variability of wave height, the largest individual waves are likely to be somewhat less than twice the reported significant wave height for a particular day or storm.

TYPES OF WIND WAVES

Three different types of wind waves develop over time:

- Capillary waves, or ripples
- Seas
- Swells

Ripples appear on smooth water when the wind blows, but will die quickly if the wind stops. The restoring force that allows them to propagate is surface tension. Seas are the larger-scale, often irregular motions that form under sustained winds. These waves tend to last much longer, even after the wind has died, and the restoring force that allows them to propagate is gravity. As waves propagate away from their area of origin, they naturally separate into groups of common direction and wavelength. The sets of waves formed in this way are known as swells.

Individual "rogue waves" (also called "freak waves", "monster waves", "killer waves", and "king waves") much higher than the other waves in the sea state can occur. In the case of the Draupner wave, its 25 m (82 ft) height was 2.2 times the significant wave height. Such waves are distinct from tides, caused by the Moon and Sun's gravitational pull, tsunamis that are caused by underwater earthquakes or landslides, and waves generated by underwater explosions or the fall of meteorites-all having far longer wavelengths than wind waves. Yet, the largest ever recorded wind waves are common-not rogue-waves in extreme sea states. For example: 29.1 m (95 ft) high waves have been recorded on the RRS Discovery in a sea with 18.5 m (61 ft) significant wave height, so the highest wave is only 1.6 times the significant wave height. The biggest recorded by a buoy (as of 2011) was 32.3 m (106 ft) high during the 2007 typhoon Krosa near Taiwan.

WIND WAVE MODELS

Surfers are very interested in the wave forecasts. There are many websites that provide predictions of the surf quality for the upcoming days and weeks. Wind wave models are driven by more general weather models that predict the winds and pressures over the oceans, seas and lakes.

Wind wave models are also an important part of examining the impact of shore protection and beach nourishment proposals. For many beach areas there is only patchy information about the wave climate, therefore estimating the effect of wind waves is important for managing littoral environments.

WAVE SHOALING

In fluid dynamics, wave shoaling is the effect by which surface waves entering shallower water increase in wave height (which is about twice the amplitude). It is caused by the fact that the group velocity, which is also the wave-energy transport velocity, decreases with the reduction of water depth. Under stationary conditions, this decrease in transport speed must be compensated by an increase in energy density in order to maintain a constant energy flux. Shoaling waves will also exhibit a reduction in wavelength while the frequency remains constant. In shallow water and parallel depth contours, non-breaking waves will increase in wave height as the wave packet enters shallower water. This is particularly evident for tsunamis as they wax in height when approaching a coastline, with devastating results.

REFRACTION

Refraction is the change in direction of a wave due to a change in its transmission medium. Refraction is essentially a surface phenomenon. The phenomenon is mainly in governance to the law of conservation of energy and momentum. Due to change of medium, the phase velocity of the wave is changed but its frequency remains constant. This is most commonly observed when a wave passes from one medium to another at any angle other than 90° or 0°. Refraction of light is the most commonly observed phenomenon, but any type of wave can refract when it interacts with a medium, for example when sound waves pass from one medium into another or when water waves move into water of a different depth. Refraction is described by Snell's law, which states that for a given pair of media and a wave with a single frequency, the ratio of the sines of the angle of incidence ?1 and angle of refraction ?2 is equivalent to the ratio of phase velocities (v1 / v2) in the two media, or equivalently, to the opposite ratio of the indices of refraction (n2 / n1):

$$\frac{\sin\theta_1}{\sin\theta_2} = \frac{v_1}{v_2} = \frac{n_2}{n_1}.$$

In general, the incident wave is partially refracted and partially reflected; the details of this behavior are described by the Fresnel equations.

EXPLANATION

In optics, refraction is a phenomenon that often occurs when waves travel from a medium with a given refractive index to a medium with another at an oblique angle. At the boundary between the media, the wave's phase velocity is altered, usually causing a change in direction. Its wavelength increases or decreases but its frequency remains constant. For example, a light ray will refract as it enters and leaves glass, assuming there is a change in refractive index. A ray traveling along the normal (perpendicular to the boundary) will change speed, but not direction. Refraction still occurs in this case. Understanding of this concept led to the invention of lenses and the refracting telescope.

Refraction can be seen when looking into a bowl of water. Air has a refractive index of about 1.0003, and water has a refractive index of about 1.3330. If a person looks at a straight object, such as a pencil or straw, which is placed at a slant, partially in the water, the object appears to bend at the water's surface. This is due to the bending of light rays as they move from the water to the air. Once the rays reach the eye, the eye traces them back as straight lines (lines of sight). The lines of sight (shown as dashed lines) intersect at a higher position than where the actual rays originated. This causes the pencil to appear higher and the water to appear shallower than it really is. The depth that the water appears to be when viewed from above is known as the apparent depth.

This is an important consideration for spearfishing from the surface because it will make the target fish appear to be in a different place, and the fisher must aim lower to catch the fish. Conversely, an object above the water has a higher apparent height when viewed from below the water. The opposite correction must be made by an archer fish. For small angles of incidence (measured from the normal, when sin θ is approximately the same as tan θ), the ratio of apparent to real depth is the ratio of the refractive indexes of air to that of water. But as the angle of incidence approaches 90o, the apparent depth approaches zero, albeit reflection increases, which limits observation at high angles of incidence. Conversely, the apparent height approaches infinity as the angle of incidence (from below) increases, but even earlier, as the angle of total internal reflection is approached, albeit the image also fades from view as this limit is approached.

The diagram on the right shows an example of refraction in water waves. Ripples travel from the left and pass over a shallower region inclined at an angle to the wavefront. The waves travel slower in the more shallow water, so the wavelength decreases and the wave bends at the boundary. The dotted line represents the normal to the boundary. The dashed line represents the original direction of the waves. This phenomenon explains why waves on a shoreline tend to strike the shore close to a perpendicular angle. As the waves travel from deep water into shallower water near the shore, they are refracted from their original direction of travel to an angle more normal to the shoreline. Refraction is also responsible for rainbows and for the splitting of white light into a rainbow-spectrum as it passes through a glass prism. Glass has a higher refractive index than air. When a beam of white light passes from air into a material having an index of refraction that varies with frequency, a phenomenon known as dispersion occurs, in which different coloured components of the white light are refracted at different angles, i.e., they bend by different amounts at the interface, so that they become separated. The different colors correspond to different frequencies.

While refraction allows for phenomena such as rainbows, it may also produce peculiar optical phenomena, such as mirages and Fata Morgana. These are caused by the change of the refractive index of air with temperature. The refractive index of materials can also be nonlinear, as occurs with the Kerr effect when high intensity light leads to a refractive index proportional to the intensity of the incident light.

Recently some metamaterials have been created which have a negative refractive index. With metamaterials, we can also obtain total refraction phenomena when the wave impedances of the two media are matched. There is then no reflected wave. Also, since refraction can make objects appear closer than they are, it is responsible for allowing water to magnify objects. First, as light is entering a drop of water, it slows down. If the water's surface is not flat, then the light will be bent into a new path. This round shape will bend the light outwards and as it spreads out, the image you see gets larger.

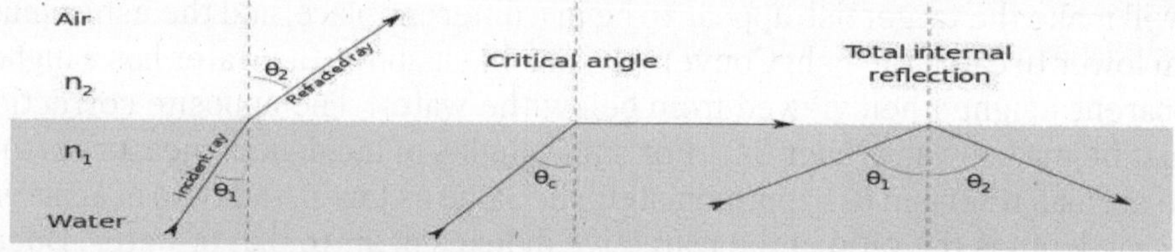

Fig. Refraction of light at the interface between two media.

A useful analogy in explaining the refraction of light would be to imagine a marching band as they march at an oblique angle from pavement (a fast medium) into mud (a slower medium). The marchers on the side that runs into the mud first will slow down first. This causes the whole band to pivot slightly toward the normal (make a smaller angle from the normal).

CLINICAL SIGNIFICANCE

In medicine, particularly optometry, ophthalmology and orthoptics, refraction (also known as refractometry) is a clinical test in which a phoropter may be used by the appropriate eye care professional to determine the eye's refractive error and the best corrective lenses to be prescribed. A series of test lenses in graded optical powers or focal lengths are presented to determine which provides the sharpest, clearest vision.

ACOUSTICS

In underwater acoustics, refraction is the bending or curving of a sound ray that results when the ray passes through a sound speed gradient from a region of one sound speed to a region of a different speed. The amount of ray bending is dependent upon the amount of difference between sound speeds, that is, the variation in temperature, salinity, and pressure of the water. Similar acoustics effects are also found in the Earth's atmosphere. The phenomenon of refraction of sound in the atmosphere has been known for centuries; however, beginning in the early 1970s, widespread analysis of this effect came into vogue through the designing of urban highways and noise barriers to address the meteorological effects of bending of sound rays in the lower atmosphere.

WIND WAVE MODEL

In fluid dynamics, wind wave modeling describes the effort to depict the sea state and predict the evolution of the energy of wind waves using numerical techniques. These simulations consider atmospheric wind forcing, nonlinear wave interactions, and frictional dissipation, and they output statistics describing wave heights, periods, and propagation directions for regional seas or global oceans. Such wave hindcasts and wave forecasts are extremely important for commercial interests on the high seas. For example, the shipping industry requires guidance for operational planning and tactical seakeeping purposes.

For the specific case of predicting wind wave statistics on the ocean, the term ocean surface wave model is used. Other applications, in particular coastal engineering, have led to the developments of wind wave models specifically designed for coastal applications.

HISTORICAL OVERVIEW

Early forecasts of the sea state were created manually based upon empirical relationships between the present state of the sea, the expected wind conditions, the fetch/duration, and the direction of the wave propagation. Alternatively, the swell part of the state has been forecasted as early as 1920 using remote observations.

During the 1950s and 1960s, much of the theoretical groundwork necessary for numerical descriptions of wave evolution was laid. For forecasting purposes, it was realized that the random nature of the sea state was best described by a spectral decomposition in which the energy of the waves was attributed to as many wave trains as necessary, each with a specific direction and period. This approach allowed to make combined forecasts of wind seas and swells. The first numerical model based on the spectral decomposition of the sea state was operated in 1956 by the French Weather Service, and focused on the North Atlantic. The 1970s saw the first operational, hemispheric wave model: the spectral wave ocean model (SWOM) at the Fleet Numerical Oceanography Center. First generation wave models did not consider nonlinear wave interactions. Second generation models, available by the early 1980s, parameterized these interactions. They included the "coupled hybrid" and "coupled discrete" formulations. Third generation models explicitly represent all the physics relevant for the development of the sea state in two dimensions. The wave modeling project (WAM), an international effort, led to the refinement of modern wave modeling techniques during the decade 1984-1994.

Improvements included two-way coupling between wind and waves, assimilation of satellite wave data, and medium-range operational forecasting.

Wind wave models are used in the context of a forecasting or hindcasting system. Differences in model results arise, with decreasing order of importance, from differences in wind and sea ice forcing, differences in parameterizations of physical processes, the use of data assimilation and associated methods, the numerical techniques used to solve the wave energy evolution equation.

GENERAL STRATEGY

Input

A wave model requires as initial conditions information describing the state of the sea. An analysis of the sea or ocean can be created through data assimilation, where observations such as buoy or satellite altimeter measurements are combined with a background guess from a previous forecast or climatology to create the best estimate of the current conditions. In practice, many forecasting system rely only on the previous forecast, without any assimilation of observations. A more critical input is the "forcing" by wind fields: a time-varying map of wind speed and directions. The most common sources of errors in wave model results are the errors in the wind field. Ocean currents can also be important, in particular in western boundary currents such as the Gulf Stream, Kuroshio or Agulhas current, or in coastal areas where tidal currents are strong. Waves are also affected by sea ice and icebergs, and all operational global wave models take at least the sea ice into account.

Representation

The sea state is described as a spectrum; the sea surface can be decomposed into waves of varying frequencies using the principle of superposition. The waves are also separated by their direction of propagation. The model domain size can range from regional to the global ocean. Smaller domains can be nested within a global domain to provide higher resolution in a region of interest. The sea state evolves according to physical equations, which include wave propagation / advection, and a source function which allows for wave energy to be augmented or diminished. The source function has at least three terms: wind forcing, nonlinear transfer, and dissipation by whitecapping. Wind data are typically provided from a separate atmospheric model from an operational weather forecasting center. For intermediate water depths the effect of bottom friction should also be added. At ocean scales, the dissipation of swells - without breaking- is a very important term.

Output

The output of a wind wave model is a description of the wave spectra, with amplitudes associated with each frequency and propagation direction. Results are typically summarized by the significant wave height, which is the average height of the one-third largest waves, and the period and propagation direction of the dominant wave.

Coupled models

Wind waves also act to modify atmospheric properties through frictional drag of near-surface winds and heat fluxes. Two-way coupled models allow the wave activity to feed back upon the atmosphere. The European Centre for Medium-Range Weather Forecasts (ECMWF) coupled atmosphere-wave forecast system described below facilitates this through exchange of the Charnock parameter which controls the sea surface roughness. This allows the atmosphere to respond to changes in the surface roughness as the wind sea builds up or decays.

EXAMPLES

WAVEWATCH

The operational wave forecasting systems at NOAA are based on the WAVEWATCH III model. This system has a global domain of approximately 100 km resolution, with nested regional domains for the northern hemisphere oceanic basins at approximately 25 km resolution. Physics includes wave field refraction, nonlinear resonant interactions, sub-grid representations of unresolved islands, and dynamically updated ice coverage. Wind data is provided from the GDAS data assimilation system for the GFS weather model. Up to 2008, the model was limited to regions outside the surf zone where the waves are not strongly impacted by shallow depths. The model can handle the effects of currents on waves from its early design by Hendrik Tolman in the 1990s, and is now extended for nearshore applications.

WAM

The wave model WAM was the first so-called third generation prognostic wave model where the two-dimensional wave spectrum was allowed to evolve freely (up to a cut-off frequency) with no constraints on the spectral shape. The model underwent a series of software updates from its inception in the late 1980s. The last official release is Cycle 4.5, maintained by the German Helmholtz Zentrum, Geesthacht.

ECMWF has incorporated WAM into its deterministic and ensemble forecasting system., known as the Integrated Forecast System (IFS). The model currently comprises 36 frequency bins and 36 propagation directions at an average spatial resolution of 25 km. The model has been coupled to the atmospheric component of IFS since 1998.

Other models

Wind wave forecasts are issued regionally by Environment Canada. Regional wave predictions are also produced by universities, such as Texas A&M University's use of the SWAN model (developed by Delft University of Technology) to forecast waves in the Gulf of Mexico. Other wind wave models include the U.S. Navy Standard Surf Model (NSSM). The private sector is also active in producing wind wave simulations and surf forecasts. For example, Oceanweather Inc. provides global operational forecasts and hindcasts of the sea state.

VALIDATION

Comparison of the wave model forecasts with observations is essential for characterizing model deficiencies and identifying areas for improvement. In-situ observations are obtained from buoys, ships and oil platforms. Altimetry data from satellites, such as GEOSAT and TOPEX, can also be used to infer the characteristics of wind waves. Hindcasts of wave models during extreme conditions also serves as a useful test bed for the models.

REANALYSES

A retrospective analysis, or reanalysis, combines all available observations with a physical model to describe the state of a system over a time period of decades. Wind waves are a part of both the NCEP Reanalysis and the ERA-40 from the ECMWF. Such resources permit the creation of monthly wave climatologies, and can track the variation of wave activity on interannual and multi-decadal time scales. During the northern hemisphere winter, the most intense wave activity is located in the central North Pacific south of the Aleutians, and in the central North Atlantic south of Iceland. During the southern hemisphere winter, intense wave activity circumscribes the pole at around 50°S, with 5 m significant wave heights typical in the southern Indian Ocean.

AIRY WAVE THEORY

In fluid dynamics, Airy wave theory (often referred to as linear wave theory) gives a linearised description of the propagation of gravity waves on the surface of a homogeneous fluid layer. The theory assumes that the fluid layer has a uniform mean depth, and that the fluid flow is inviscid, incompressible and irrotational. This theory was first published, in correct form, by George Biddell Airy in the 19th century. Airy wave theory is often applied in ocean engineering and coastal engineering for the modelling of random sea states - giving a description of the wave kinematics and dynamics of high-enough accuracy for many purposes. Further, several second-order nonlinear properties of surface gravity waves, and their propagation, can be estimated from its results. Airy wave theory is also a good approximation for tsunami waves in the ocean, before they steepen near the coast. This linear theory is often used to get a quick and rough estimate of wave characteristics and their effects. This approximation is accurate for small ratios of the wave height to water depth (for waves in shallow water), and wave height to wavelength (for waves in deep water).

DESCRIPTION

Airy wave theory uses a potential flow (or velocity potential) approach to describe the motion of gravity waves on a fluid surface. The use of - inviscid and irrotational - potential flow in water waves is remarkably successful, given its failure to describe many other fluid flows where it is often essential to take viscosity, vorticity, turbulence and/or flow separation into account. This is due to the fact that for the oscillatory part of the fluid motion, wave-induced vorticity is restricted to some thin oscillatory Stokes boundary layers at the boundaries of the fluid domain. Airy wave theory is often used in ocean engineering and coastal engineering. Especially for random waves, sometimes called wave turbulence, the evolution of the wave statistics - including the wave spectrum - is predicted well over not too long distances (in terms of wavelengths) and in not too shallow water. Diffraction is one of the wave effects which can be described with Airy wave theory. Further, by using the WKBJ approximation, wave shoaling and refraction can be predicted.

Earlier attempts to describe surface gravity waves using potential flow were made by, among others, Laplace, Poisson, Cauchy and Kelland. But Airy was the first to publish the correct derivation and formulation in 1841. Soon after, in 1847, the linear theory of Airy was extended by Stokes for non-linear wave motion - known as Stokes' wave theory - correct up to third order in the wave steepness. Even before Airy's linear theory, Gerstner derived a nonlinear trochoidal wave theory in 1804, which however is not irrotational.

Airy wave theory is a linear theory for the propagation of waves on the surface of a potential flow and above a horizontal bottom. The free surface elevation $\eta(x,t)$ of one wave component is sinusoidal, as a function of horizontal position x and time t:

$$\eta(x,t)=a\cos(kx-\omega t)$$

where

- a is the wave amplitude in metre,
- cos is the cosine function,
- k is the angular wavenumber in radian per metre, related to the wavelength λ as

$$k=\frac{2\pi}{\lambda},$$

- ω is the angular frequency in radian per second, related to the period T and frequency f by

$$\omega=\frac{2\pi}{T}=2\pi f.$$

The waves propagate along the water surface with the phase speed c_p:

$$c_p=\frac{\omega}{k}=\frac{\lambda}{T}.$$

The angular wavenumber k and frequency ω are not independent parameters (and thus also wavelength λ and period T are not independent), but are coupled. Surface gravity waves on a fluid are dispersive waves - exhibiting frequency dispersion - meaning that each wavenumber has its own frequency and phase speed.

Note that in engineering the wave height H - the difference in elevation between crest and trough - is often used:

$$H=2a \quad \text{and} \quad a=\frac{1}{2}H,$$

valid in the present case of linear periodic waves.

Underneath the surface, there is a fluid motion associated with the free surface motion. While the surface elevation shows a propagating wave, the fluid particles are in an orbital motion. Within the framework of Airy wave theory, the orbits are closed curves: circles in deep water, and ellipses in finite depth- with the ellipses becoming flatter near the bottom of the fluid layer. So while the wave propagates, the fluid particles just orbit (oscillate) around their average position. With the propagating wave motion, the fluid particles transfer energy in the wave propagation direction, without having a mean velocity. The diameter of the orbits reduces with depth below the free surface. In deep water, the orbit's diameter is reduced to 4% of its free-surface value at a depth of half a wavelength.

In a similar fashion, there is also a pressure oscillation underneath the free surface, with wave-induced pressure oscillations reducing with depth below the free surface - in the same way as for the orbital motion of fluid parcels.

MATHEMATICAL FORMULATION OF THE WAVE MOTION

FLOW PROBLEM FORMULATION

The waves propagate in the horizontal direction, with coordinate x, and a fluid domain bound above by a free surface at z = η(x,t), with z the vertical coordinate (positive in the upward direction) and t being time. The level z = 0 corresponds with the mean surface elevation.

The impermeable bed underneath the fluid layer is at z = -h. Further, the flow is assumed to be incompressible and irrotational - a good approximation of the flow in the fluid interior for waves on a liquid surface - and potential theory can be used to describe the flow. The velocity potential Φ (x,z,t) is related to the flow velocity components u_x and u_z in the horizontal (x) and vertical (z) directions by:

$$u_x=\frac{\partial\Phi}{\partial x} \quad \text{and} \quad u_z=\frac{\partial\Phi}{\partial z}.$$

Then, due to the continuity equation for an incompressible flow, the potential Φ has to satisfy the Laplace equation:

$$\frac{\partial^2\Phi}{\partial x^2}+\frac{\partial^2\Phi}{\partial z^2}=0.$$

Boundary conditions are needed at the bed and the free surface in order to close the system of equations. For their formulation within the framework of linear theory, it is necessary to specify what the base state (or zeroth-order solution) of the flow is. Here, we assume the base state is rest, implying the mean flow velocities are zero. The bed being impermeable, leads to the kinematic bed boundary-condition:

$$\frac{\partial\Phi}{\partial z}=0 \quad \text{at } z=-h.$$

In case of deep water - by which is meant infinite water depth, from a mathematical point of view - the flow velocities have to go to zero in the limit as the vertical coordinate goes to minus infinity: $z \to -\infty$.

At the free surface, for infinitesimal waves, the vertical motion of the flow has to be equal to the vertical velocity of the free surface. This leads to the kinematic free-surface boundary-condition:

$$\frac{\partial\eta}{\partial t}=\frac{\partial\Phi}{\partial z} \quad \text{at } z=\eta(x,t).$$

If the free surface elevation $\eta(x,t)$ was a known function, this would be enough to solve the flow problem. However, the surface elevation is an extra unknown, for which an additional boundary condition is needed. This is provided by Bernoulli's equation for an unsteady potential flow. The pressure above the free surface is assumed to be constant. This constant pressure is taken equal to zero, without loss of generality, since the level of such a constant pressure does not alter the flow. After linearisation, this gives the dynamic free-surface boundary condition:

$$\frac{\partial\Phi}{\partial t}+g\eta=0 \quad \text{at } z=\eta(x,t).$$

Because this is a linear theory, in both free-surface boundary conditions - the kinematic and the dynamic one, equations (3) and (4) - the value of Φ and $\partial\Phi/\partial z$ at the fixed mean level $z = 0$ is used.

TABLE OF WAVE QUANTITIES

In the table below, several flow quantities and parameters according to

Airy wave theory are given. The given quantities are for a bit more general situation as for the solution given above. Firstly, the waves may propagate in an arbitrary horizontal direction in the x = (x,y) plane. The wavenumber vector is k, and is perpendicular to the cams of the wave crests. Secondly, allowance is made for a mean flow velocity U, in the horizontal direction and uniform over (independent of) depth z.

This introduces a Doppler shift in the dispersion relations. At an Earth-fixed location, the observed angular frequency (or absolute angular frequency) is ω. On the other hand, in a frame of reference moving with the mean velocity U (so the mean velocity as observed from this reference frame is zero), the angular frequency is different. It is called the intrinsic angular frequency (or relative angular frequency), denoted as σ. So in pure wave motion, with U=0, both frequencies ω and σ are equal. The wave number k (and wavelength λ) are independent of the frame of reference, and have no Doppler shift (for monochromatic waves). The table only gives the oscillatory parts of flow quantities - velocities, particle excursions and pressure - and not their mean value or drift. The oscillatory particle excursions ξ_x and ξ_z are the time integrals of the oscillatory flow velocities ux and uz respectively.

Water depth is classified into three regimes:

- deep water - for a water depth larger than half the wavelength, h>½λ, the phase speed of the waves is hardly influenced by depth (this is the case for most wind waves on the sea and ocean surface),
- shallow water - for a water depth smaller than the wavelength divided by 20, h<½20 λ, the phase speed of the waves is only dependent on water depth, and no longer a function of period or wavelength; and
- intermediate depth - all other cases, $/_{20}$ λ<h < ½ λ, where both water depth and period (or wavelength) have a significant influence on the solution of Airy wave theory.

In the limiting cases of deep and shallow water, simplifying approximations to the solution can be made. While for intermediate depth, the full formulations have to be used.

TWO LAYERS BOUNDED ABOVE BY A FREE SURFACE

In this case the dispersion relation allows for two modes: a barotropic mode where the free surface amplitude is large compared with the amplitude of the interfacial wave, and a baroclinic mode where the opposite is the case - the interfacial wave is higher than and in antiphase with the free surface wave. The dispersion relation for this case is of a more complicated form.

SECOND-ORDER WAVE PROPERTIES

Several second-order wave properties, i.e. quadratic in the wave amplitude a, can be derived directly from Airy wave theory. They are of importance in many practical applications, e.g. forecasts of wave conditions. Using a WKBJ approximation, second-order wave properties also find their applications in describing waves in case of slowly varying bathymetry, and mean-flow variations of currents and surface elevation. As well as in the description of the wave and mean-flow interactions due to time and space-variations in amplitude, frequency, wavelength and direction of the wave field itself.

BREAKING WAVE

In fluid dynamics, a breaking wave is a wave whose amplitude reaches a critical level at which some process can suddenly start to occur that causes large amounts of wave energy to be transformed into turbulent kinetic energy. At this point, simple physical models that describe wave dynamics often become invalid, particularly those that assume linear behaviour. The most generally familiar sort of breaking wave is the breaking of water surface waves on a coastline. Because of the horizontal component of the fluid velocity associated with the wave motion, wave crests steepen as the amplitude increases; wave breaking generally occurs where the amplitude reaches the point that the crest of the wave actually overturns-though the types of breaking water surface waves are discussed in more detail below. Certain other effects in fluid dynamics have also been termed "breaking waves," partly by analogy with water surface waves. In meteorology, atmospheric gravity waves are said to break when the wave produces regions where the potential temperature decreases with height, leading to energy dissipation through convective instability; likewise Rossby waves are said to break when the potential vorticity gradient is overturned. Wave breaking also occurs in plasmas, when the particle velocities exceed the wave's phase speed.

BREAKING WATER SURFACE WAVES

Breaking of water surface waves may occur anywhere that the amplitude is sufficient, including in mid-ocean. However, it is particularly common on beaches because wave heights are amplified in the region of shallower water (because the group velocity is lower there). See also waves and shallow water.

There are four basic types of breaking water waves. They are spilling, plunging, collapsing, and surging.

Spilling breakers

When the ocean floor has a gradual slope, the wave will steepen until the crest becomes unstable, resulting in turbulent whitewater spilling down the face of the wave. This continues as the wave approaches the shore, and the wave's energy is slowly dissipated in the whitewater. Because of this, spilling waves break for a longer time than other waves, and create a relatively gentle wave. Onshore wind conditions make spillers more likely.

Plunging breakers

A plunging wave occurs when the ocean floor is steep or has sudden depth changes, such as from a reef or sandbar. The crest of the wave becomes much steeper than a spilling wave, becomes vertical, then curls over and drops onto the trough of the wave, releasing most of its energy at once in a relatively violent impact. A plunging wave breaks with more energy than a significantly larger spilling wave. The wave can trap and compress the air under the lip, which creates the "crashing" sound associated with waves. With large waves, this crash can be felt by beachgoers on land. Offshore wind conditions can make plungers more likely.

If a plunging wave is not parallel to the beach (or the ocean floor), the section of the wave which reaches shallow water will break first, and the breaking section (or curl) will move laterally across the face of the wave as the wave continues. This is the "tube" that is so highly sought after by surfers (also called a "barrel", a "pit", and "the greenroom", among other terms). The surfer tries to stay near or under the curling lip, often trying to stay as "deep" in the tube as possible while still being able to shoot forward and exit the tube before it closes. A plunging wave that is parallel to the beach can break along its whole length at once, rendering it unrideable and dangerous. Surfers refer to these waves as "closed out".

Collapsing

Collapsing waves are a cross between plunging and surging, in which the crest never fully breaks, yet the bottom face of the wave gets steeper and collapses, resulting in foam.

Surging

Surging breakers originate from long period, low steepness waves and/or steep beach profiles. The outcome is the rapid movement of the base of the wave up the swash slope and the disappearance of the wave crest. The front face and crest of the wave remain relatively smooth with little foam or bubbles, resulting in a very narrow surf zone, or no breaking waves at all. The short, sharp burst of wave energy means that the swash/backwash cycle completes before the arrival of the next wave, leading to a low value of Kemp's phase difference (< 0.5). Surging waves are typical of reflective beach states. On steeper beaches, the energy of the wave can be reflected by the bottom back into the ocean, causing standing waves.

PHYSICS

During breaking, a deformation (usually a bulge) forms at the wave crest, either leading side of which is known as the "toe." Parasitic capillary waves are formed, with short wavelengths. Those above the "toe" tend to have much longer wavelengths. This theory is anything but perfect, however, as it's linear. There have been a couple non-linear theories of motion (regarding waves). One put forth uses a perturbation method to expand the description all the way to the third order, and better solutions have been found since then. As for wave deformation, methods much like the boundary integral method and the Boussinesq model have been created.

It has been found that high-frequency detail present in a breaking wave plays a part in crest deformation and destabilization. The same theory expands on this, stating that the valleys of the capillary waves create a source for vorticity. It is said that surface tension (and viscosity) are significant for waves up to 2m in wavelength.

These models are flawed, however, as they can't take into account what happens to the water after the wave breaks. Post-break eddy forms and the turbulence created via the breaking is mostly unresearched. Understandably, it might be difficult to glean predictable results from the ocean. After the tip of the wave overturns and the jet collapses, it creates a very coherent and defined horizontal vertex. The plunging breakers create secondary eddies down the face of the wave. Small horizontal random eddides that form on the sides of the wave suggest that, perhaps, prior to breaking, the water's velocity is more or less two dimensional. This becomes three dimensional upon breaking.

The main vortex along the front of the wave diffuses rapidly into the interior of the wave after breaking, as the eddies on the surface become more viscous. Advection and molecular diffusion play a part in stretching the vortex and redistributing the vorticity, as well as the formation turbulence cascades. The energy of the large vortices are, by this method, is transferred to much smaller isotropic vortices. Experiments have been conducted to deduce the evolution of turbulence after break, both in deep water and on a beach.

3

Electric Power

An electric power system is a network of electrical components used to supply, transmit and use electric power. An example of an electric power system is the network that supplies a region's homes and industry with power - for sizable regions, this power system is known as *the grid* and can be broadly divided into the generators that supply the power, the transmission system that carries the power from the generating centres to the load centres and the distribution system that feeds the power to nearby homes and industries. Smaller power systems are also found in industry, hospitals, commercial buildings and homes.

The majority of these systems rely upon three-phase AC power - the standard for large-scale power transmission and distribution across the modern world. Specialised power systems that do not always rely upon three-phase AC power are found in aircraft, electric rail systems, ocean liners and automobiles.

Fig. A steam turbine used to provide electric power.

HISTORY

In 1881 two electricians built the world's first power system at Godalming in England. It was powered by a power station consisting of two waterwheels that produced an alternating current that in turn supplied seven Siemens arc lamps at 250 volts and 34 incandescent lamps at 40 volts. However supply to the lamps was intermittent and in 1882 Thomas Edison and his company, The Edison Electric Light Company, developed the first steam powered electric power station on Pearl Street in New York City. The Pearl Street Station initially powered around 3,000 lamps for 59 customers. The power station used direct current and operated at a single voltage. Direct current power could not be easily transformed to the higher voltages necessary to minimise power loss during long-distance transmission, so the maximum economic distance between the generators and load was limited to around half-a-mile (800 m).

That same year in London Lucien Gaulard and John Dixon Gibbs demonstrated the first transformer suitable for use in a real power system. The practical value of Gaulard and Gibbs' transformer was demonstrated in 1884 at Turin where the transformer was used to light up forty kilometres (25 miles) of railway from a single alternating current generator. Despite the success of the system, the pair made some fundamental mistakes. Perhaps the most serious was connecting the primaries of the transformers in series so that active lamps would affect the brightness of other lamps further down the line. Following the demonstration George Westinghouse, an American entrepreneur, imported a number of the transformers along with a Siemens generator and set his engineers to experimenting with them in the hopes of improving them for use in a commercial power system. In July 1888, Westinghouse also licensed Nikola Tesla's US patents for a polyphase AC induction motor and transformer designs and hired Tesla for one year to be a consultant at the Westinghouse Electric and Manufacturing Company's Pittsburgh labs.

One of Westinghouse's engineers, William Stanley, recognised the problem with connecting transformers in series as opposed to paralleland also realised that making the iron core of a transformer a fully enclosed loop would improve the voltage regulation of the secondary winding. Using this knowledge he built a much improved alternating current power system at Great Barrington, Massachusetts in 1886.

By 1890 the electric power industry was flourishing, and power companies had built thousands of power systems (both direct and alternating current) in the United States and Europe. These networks were effectively dedicated to providing electric lighting. During this time a fierce rivalry known as the "War of Currents" emerged between Thomas Edison and George Westinghouse over which form of transmission (direct or alternating current) was superior. In 1891, Westinghouse installed the first major power system that was designed to drive a 100 horsepower (75 kW) synchronous electric motor, not just provide electric lighting, at Telluride, Colorado. On the other side of the Atlantic,Oskar von

Miller built a 20 kV 176 km three-phase transmission line from Lauffen am Neckar to Frankfurt am Main for the Electrical Engineering Exhibition in Frankfurt. In 1895, after a protracted decision-making process, the Adams No. 1 generating station at Niagara Falls began transferring three-phase alternating current power to Buffalo at 11 kV. Following completion of the Niagara Falls project, new power systems increasingly chose alternating current as opposed to direct current for electrical transmission.

Developments in power systems continued beyond the nineteenth century. In 1936 the first experimental HVDC (high voltage direct current) line using mercury arc valves was built between Schenectady and Mechanicville, New York. HVDC had previously been achieved by series-connected direct current generators and motors (the Thury system) although this suffered from serious reliability issues.

In 1957 Siemens demonstrated the first solid-state rectifier, but it was not until the early 1970s that solid-state devices became the standard in HVDC. In recent times, many important developments have come from extending innovations in the ICT field to the power engineering field. For example, the development of computers meant load flow studies could be run more efficiently allowing for much better planning of power systems. Advances in information technology and telecommunication also allowed for remote control of a power system's switchgear and generators.

ELECTRIC POWER

Fig. Electric power is transmitted onoverhead lines like these, and also on underground high voltage cables.

Electric power is the rate at which electric energy is transferred by an electric circuit. The SI unit of power is the watt, one joule persecond. Electric power is usually produced by electric generators, but can also be supplied by sources such as electric batteries. It is generally supplied to businesses and homes by the electric power industry through an electric power grid. Electric power is usually sold by thekilowatt hour (3.6 MJ) which is the product of power in kilowatts multiplied by running time in hours. Electric utilities measure power using an electricity meter, which keeps a running total of the electric energy delivered to a customer.

DEFINITION

Electric power, like mechanical power, is the rate of doing work, measured in watts, and represented by the letter *P*. The term *wattage* is used colloquially to mean "electric power in watts." The electric power in watts produced by an electric current *I* consisting of a charge of *Q* coulombs every *t* seconds passing through an electric potential (voltage) difference of *V* is

$$P = \text{work done per unit time} = \frac{VQ}{t} = VI$$

where
Q is electric charge in coulombs
t is time in seconds
I is electric current in amperes
V is electric potential or voltage in volts

ELECTRIC POWER SUPPLY

ELECTRICITY GENERATION

The fundamental principles of electricity generation were discovered during the 1820s and early 1830s by the British scientist Michael Faraday. His basic method is still used today: electricity is generated by the movement of a loop of wire, or disc of copper between the poles of a magnet.

For electric utilities, it is the first process in the delivery of electricity to consumers. The other processes, electricity transmission, distribution, and electrical power storage and recovery using pumped-storage methods are normally carried out by the electric power industry.

Electricity is most often generated at a power station by electromechanical generators, primarily driven by heat engines fueled by chemical combustion or nuclear fission but also by other means such as the kinetic energy of flowing water and wind. There are many other technologies that can be and are used to generate electricity such as solarphotovoltaics and geothermal power.

BATTERY POWER

A battery is a device consisting of one or more electrochemical cells that convert stored chemical energy into electrical energy. Since the invention of the first battery (or "voltaic pile") in 1800 by Alessandro Volta and especially since the technically improved Daniell cell in 1836, batteries have become a common power source for many household and industrial applications. According to a 2005 estimate, the worldwide battery industry generates US$48 billion in sales each year, with 6per cent annual growth.

There are two types of batteries: primary batteries (disposable batteries), which are designed to be used once and discarded, and secondary batteries (rechargeable batteries), which are designed to be recharged and used multiple times. Batteries come in many sizes, from miniature cells used to power hearing aids and wristwatches to battery banks the size of rooms that provide standby power for telephone exchanges and computer data centers.

ELECTRIC POWER INDUSTRY

The electric power industry provides the production and delivery of power, in sufficient quantities to areas that need electricity, through a grid connection. The grid distributes electrical energy to customers. Electric power is generated by central power stations or by distributed generation.

Many households and businesses need access to electricity, especially in developed nations, the demand being scarcer in developing nations. Demand for electricity is derived from the requirement for electricity in order to operate domestic appliances, office equipment, industrial machinery and provide sufficient energy for both domestic and commercial lighting, heating, cooking and industrial processes. Because of this aspect of the industry, it is viewed as a public utility as infrastructure.

DEFINITION OF ELECTRIC POWER

The electric power P is equal to the energy consumption E divided by the consumption time t:

$$P = \frac{E}{t}$$

P is the electric power in watt (W).

E is the energy consumption in joule (J).

t is the time in seconds (s).

Example

Find the electric power of an electrical circuit that consumes 120 joules for 20 seconds.

Solution:

$E = 120\text{J}$

$t = 20\text{s}$

$P = E/t = 120\text{J}/20\text{s} = 6\text{W}$

Electric power calculation

$P = V \cdot I$

or

$P = I^2 \cdot R$

or

$P = V^2 / R$

P is the electric power in watt (W).

V is the voltage in volts (V).

I is the current in amps (A).

R is the resistance in ohms (Ù).

POWER OF AC CIRCUITS

The formulas are for single phase AC power.

For 3 phase AC power:

When line to line voltage (V_{L-L}) is used in the formula, multiply the single phase power by square root of 3 (“3=1.73).

When line to zero voltage (V_{L-0}) is used in the formula, multiply the single phase power by 3.

REAL POWER

Real or true power is the power that is used to do the work on the load.

$P = V_{rms} I_{rms} \cos ö$

P is the real power in watts [W]

V_{rms} is the rms voltage = V_{peak}/”2 in Volts [V]

I_{rms} is the rms current = I_{peak}/”2 in Amperes [A]

ö is the impedance phase angle = phase difference between voltage and current.

REACTIVE POWER

Reactive power is the power that is wasted and not used to do work on the load.

$Q = V_{rms} I_{rms} \sin ö$

Q is the reactive power in volt-ampere-reactive [VAR]

V_{rms} is the rms voltage = V_{peak}/”2 in Volts [V]

I_{rms} is the rms current = I_{peak}/”2 in Amperes [A]

ö is the impedance phase angle = phase difference between voltage and current.

APPARENT POWER

The apparent power is the power that is supplied to the circuit.

$S = V_{rms} I_{rms}$

S is the apparent power in Volt-ampér [VA]
V_{rms} is the rms voltage = V_{peak}/"2 in Volts [V]
I_{rms} is the rms current = I_{peak}/"2 in Amperes [A]

REAL/ REACTIVE/ APPARENT POWERS RELATION

The real power P and reactive power Q give together the apparent power S:

$P^2 + Q^2 = S^2$

P is the real power in watts [W]
Q is the reactive power in volt-ampere-reactive [VAR]
S is the apparent power in Volt-amper [VA]

ELECTRONICS BASICS: ELECTRICAL POWER

The three key concepts you need to know before working with electronic circuits are current, voltage, and power. *Current* is the organized flow of electric charges through a conductor, and *voltage* is the driving force that pushes electric charges to create current.

The third piece of the puzzle is called *power* (abbreviated P in equations). Simply put, power is the work done by an electric circuit.

Electric current, in and of itself, isn't all that useful. It becomes useful only when the energy carried by an electric current is converted into some other form of energy, such as heat, light, sound, or radio waves. For example, in an incandescent light bulb, voltage pushes current through a *filament*, which converts the energy carried by the current into heat and light.

Power is measured in units called *watts* (abbreviated *W*). The definition of one watt is simple: One watt is the amount of work done by a circuit in which one ampere of current is driven by one volt.

This relationship lends itself to a simple equation. Fortunately, this one is pretty simple:

$$P = E \times I$$

In other words, *power* (P) equals *voltage* (E) times *current* (I).

To use the equation correctly, you must make sure that you measure power, voltage, and current using their standard units: watts, volts, and amperes. For example, suppose you have a light bulb connected to a 10-volt power supply, and one-tenth of an ampere is flowing through the light bulb. To calculate the wattage of the light bulb, you use the P = E × I formula like this:

$$P = 10\ V \times 0.1\ A = 1\ W$$

Thus, the light bulb is doing 1 watt of work.

Often, you know the voltage and the wattage of the circuit and you want to use those values to determine the amount of current flowing through the circuit. You can do that by turning the equation around, like this:

$$I = \frac{P}{E}$$

For example, if you want to determine how much current flows through a lamp with a 100-watt light bulb when it's plugged into a 117-volt electrical outlet, use the formula like this:

$$\mathrm{I} = \frac{100\ \mathrm{W}}{117\ \mathrm{V}} = 0.855\ \mathrm{A}$$

Thus, the current through the circuit is 0.855 amperes.

Here are some final thoughts concerning the concept of power:

- The term *dissipate* is often used in association with power. As the energy carried by an electric current is converted into another form such as heat or light, the circuit is said to dissipate power.
- Did you notice that current and voltage are represented by the letters I and E, not the letters C or V as you might expect, but power is represented by the letter P? Sometimes you wonder if the people who make the rules are just trying to confuse everyone.

Maybe the following table will help you keep things sorted out:

Concept	Abbreviation	Unit
Current	I	amp (A)
Voltage	E or EMF	volt (V)
Power	P	watt (W)

- The definition of a volt is simple: One volt is the amount of electromotive force (EMF) necessary to do one watt of work at one ampere of current.
- The P = E × I formula is sometimes called *Joule's Law*, named after the person who discovered it.
- Calculating the power dissipated by a circuit is often a very important part of circuit design. That's because electrical components such as resistors, transistors, capacitors, and integrated circuits all have maximum power ratings.

POWER IN ELECTRIC CIRCUITS

In addition to voltage and current, there is another measure of free electron activity in a circuit: *power*. First, we need to understand just what power is before we analyze it in any circuits.

Power is a measure of how much work can be performed in a given amount of time. *Work* is generally defined in terms of the lifting of a weight against the pull of gravity. The heavier the weight and/or the higher it is lifted, the more work has been done. *Power* is a measure of how rapidly a standard amount of work is done.

For American automobiles, engine power is rated in a unit called "horsepower," invented initially as a way for steam engine manufacturers to quantify the working ability of their machines in terms of the most common power source of their day: horses. One horsepower is defined in British units as 550 ft-lbs of work per second of time. The power of a car's engine won't indicate how tall of a hill it can climb or how much weight it can tow, but it will indicate how *fast* it can climb a specific hill or tow a specific weight.

The power of a mechanical engine is a function of both the engine's speed and its torque provided at the output shaft. Speed of an engine's output shaft is measured in revolutions per minute, or RPM. Torque is the amount of twisting force produced by the engine, and it is usually measured in pound-feet, or lb-ft (not to be confused with foot-pounds or ft-lbs, which is the unit for work). Neither speed nor torque alone is a measure of an engine's power.

A 100 horsepower diesel tractor engine will turn relatively slowly, but provide great amounts of torque. A 100 horsepower motorcycle engine will turn very fast, but provide relatively little torque. Both will produce 100 horsepower, but at different speeds and different torques. The equation for shaft horsepower is simple:

$$\text{Horsepower} = \frac{2\pi S T}{33{,}000}$$

Where,

S = shaft speed in r.p.m.

T = shaft torque in lb-ft.

Notice how there are only two variable terms on the right-hand side of the equation, S and T. All the other terms on that side are constant: 2, pi, and 33,000 are all constants (they do not change in value). The horsepower varies only with changes in speed and torque, nothing else. We can re-write the equation to show this relationship:

$$\text{Horsepower} \propto S\,T$$

$\propto$ This symbol means "proportional to"

Because the unit of the "horsepower" doesn't coincide exactly with speed in revolutions per minute multiplied by torque in pound-feet, we can't say that horsepower *equals* ST. However, they are *proportional* to one another. As the mathematical product of ST changes, the value for horsepower will change by the same proportion. In electric circuits, power is a function of both voltage and current. Not surprisingly, this relationship bears striking resemblance to the "proportional" horsepower formula above:

$$P = I E$$

In this case, however, power (P) is exactly equal to current (I) multiplied by voltage (E), rather than merely being proportional to IE. When using this formula, the unit of measurement for power is the *watt*, abbreviated with the letter "W." It must be understood that neither voltage nor current by themselves constitute power. Rather, power is the combination of both voltage*and* current in a circuit.

Remember that voltage is the specific work (or potential energy) per unit charge, while current is the rate at which electric charges move through a conductor.

Voltage (specific work) is analogous to the work done in lifting a weight against the pull of gravity. Current (rate) is analogous to the speed at which that weight is lifted. Together as a product (multiplication), voltage (work) and current (rate) constitute power.

Just as in the case of the diesel tractor engine and the motorcycle engine, a circuit with high voltage and low current may be dissipating the same amount of power as a circuit with low voltage and high current.

Neither the amount of voltage alone nor the amount of current alone indicates the amount of power in an electric circuit.

In an open circuit, where voltage is present between the terminals of the source and there is zero current, there is *zero* power dissipated, no matter how great that voltage may be.

Since P=IE and I=0 and anything multiplied by zero is zero, the power dissipated in any open circuit must be zero. Likewise, if we were to have a short circuit constructed of a loop of superconducting wire (absolutely zero resistance), we could have a condition of current in the loop with zero voltage, and likewise no power would be dissipated. Since P=IE and E=0 and anything multiplied by zero is zero, the power dissipated in a superconducting loop must be zero.

Whether we measure power in the unit of "horsepower" or the unit of "watt," we're still talking about the same thing: how much work can be done in a given amount of time.

The two units are not numerically equal, but they express the same kind of thing. In fact, European automobile manufacturers typically advertise their engine power in terms of kilowatts (kW), or thousands of watts, instead of horsepower!

These two units of power are related to each other by a simple conversion formula:

1 Horsepower = 745.7 Watts

So, our 100 horsepower diesel and motorcycle engines could also be rated as "74570 watt" engines, or more properly, as "74.57 kilowatt" engines. In

European engineering specifications, this rating would be the norm rather than the exception.

SINGLE PHASE ELECTRIC POWER

The generation of AC electric power is commonly *three phase*, in which the waveforms of three supply conductors are offset from one another by 120°. The design of the power generators has three sets of coils placed 120 degrees apart rotating in a magnetic field. This creates three separate sine waves of electricity that are displaced from each other in time by 120 degrees of rotation (1/3 of a circle). Standard frequencies of rotation are either 50 Hertz (cycles per second) in Europe or 60 Hertz in North America. The voltage across any pair of these three conductors, or between a single conductor and ground (in a grounded system) is what is known as "single phase" electric power. *Single phase*power is what is commonly available to residential and light-commercial consumers in most distribution power grids. In North America, the single phase that is supplied is developed across a transformer coil at the utility pole (for aerial drop) or transformer pad (for underground) distribution. This single coil is center tapped and the tap is grounded to develop two waveforms that are 180 degrees out of phase with each other with 1/2 the voltage. This then creates a 120/240 volt system that is delivered to the customer. The voltage from either side of the coil to the center tap (ground) is 120 volts whereas the voltage between the two conductors on either end of the coil develops the full voltage of 240 volts.

INVERTERS AND BATTERY BASED AC

An *inverter* is a circuit for converting direct current to alternating current. An inverter can have one or two switched-mode power supplies (SMPS).

Early inverters consisted of an oscillator driving a transistor as an on/off switch, that is used to interrupt the incoming direct current to create a square wave. This is then fed through a transformer to smooth the square wave into a sine wave and to produce the required output voltage.

More efficient inverters use various methods to produce an approximate sine wave at the transformer input rather than relying on the transformer to smooth it. Capacitors can be used to smooth the flow of current into and out of the transformer. It is also possible to produce a more sinusoidal wave by having split-rail direct current inputs at two voltages, (positive and negative inputs with a central ground). By connecting the transformer input terminals in a timed sequence between the positive rail and ground, the positive rail and the negative rail, the ground rail and the negative rail, then both to the ground rail, a 'stepped sinusoid' is generated at the transformer input and the current drain on the direct current supply is less variable.

Modified Sine Wave inverters convert the (usually 12 V DC) battery voltage to high frequency (20 kHz) AC, so that a smaller transformer can be used for stepping up to a higher voltage (say 160 V) AC. This output is converted to DC at the same voltage, and then inverted again to a quasi sine wave output (about 120 V RMS). A disadvantage of the modified sine wave inverters is that the output voltage depends on the battery voltage.

It is quite difficult to obtain a good sine wave from an inverter. The quoted accuracy (*harmonic distortion*) for most is less than 60per cent, and will have an effect on the appliances connected to the output of the inverter. This *might* mean noisy operation in some appliances and/or damaged electric motors, because they will run less efficiently and could overheat. High end inverters (> $2,000) produce waveforms which are closer to the sine wave produced by a utility.

BATTERIES

Most home systems use conventional lead acid batteries for storage. They are cheap, and are deep cycle batteries, *i.e.*, they can be discharged completely and charged again many times. You cannot use automobile batteries in inverters, as they are only used to provide a large starting current, and are not meant to be discharged completely. The lead acid batteries have the disadvantage that they have to be replenished with distilled water every few months, and if it dries out, it cannot be repaired. However, they can provide the large surge currents which are required by many loads (such as induction motors) which may be connected to the system.

SWITCHED MODE POWER SUPPLY

Switched-mode power supplies may be designed to convert from alternating current or direct current, or both. They generally output direct current, although an inverter is technically a switched-mode power supply.

Switched-mode power supplies operate by using an inverter to convert the input direct current supply to alternating current, usually at around 20 kHz. If the input is alternating current but at a lower frequency (such as 50 Hz or 60 Hz line power) then an inverter is still used to bump the frequency up.

This high frequency means that the output transformer of the inverter will operate more efficiently than if it were run at 50 Hz or 60 Hz, due to hysteresis in the transformer core, and the transformer will not need to be as large or heavy. This high-frequency output is then fed through a rectifier to produce the output direct current. Regulation is achieved through feedback. The output voltage is compared to a reference voltage and the result used to alter the switching frequency or duty cycle of the inverter oscillator, which affects its output voltage. Switched-mode PSUs in domestic products such as personal computers often have universal inputs, meaning that they can accept power from most mains supplies throughout the world, with frequencies from 50 Hz to 60 Hz and voltages from 100 V to 240 V.

Unlike most other appliances, switched mode power supplies tend to be constant power devices, drawing more current as the line voltage reduces. This may cause stability problems in some situations such as emergency generator systems.

Also, maximum current draw occurs at the peaks of the waveform cycle. This means that basic switched mode power supplies tend to produce more harmonics and have a worse power factor than other types of appliances. However, higher-quality switched-mode power supplies with power-factor correction (PFC) are available, which are designed to present close to a resistive load to the mains.

The term power factor with respect to switched-mode supplies is misleading as it doesn't have much to do with leading or lagging voltage, but the way in which it loads the circuit (*i.e.* only at certain points in the cycle).

There are several types of switched-mode power supplies, classified according to the circuit topology.

1. buck regulator (single inductor; output voltage < input voltage)
2. boost regulator (single inductor; output voltage > input voltage)
3. buckboost regulator (single inductor; output voltage can be more or less than the input voltage)
4. flyback regulator (uses output transformer; allows multiple outputs and input-to-output isolation)
5. forward regulator (uses output transformer; allows multiple outputs and input-to-output isolation)
6. Cuk converter (uses a capacitor for energy storage; produces negative voltage for positive input)

STRAY CAPACITANCE AND INDUCTANCE

Aside from power ratings and power losses, transformers often Harbour other undesirable limitations which circuit designers must be made aware of. Like their simpler counterparts -- inductors -- transformers exhibit capacitance due to the insulation dielectric between conductors: from winding to winding, turn to turn (in a single winding), and winding to core. Usually this capacitance is of no concern in a power application, but small signal applications (especially those of high frequency) may not tolerate this quirk well.

Also, the effect of having capacitance along with the windings' designed inductance gives transformers the ability to resonate at a particular frequency, definitely a design concern in signal applications where the applied frequency may reach this point (usually the resonant frequency of a power transformer is well beyond the frequency of the AC power it was designed to operate on). Flux containment (making sure a transformer's magnetic flux doesn't escape so as to interfere with another device, and making sure other devices' magnetic flux is shielded from the transformer core) is another concern shared both by inductors and transformers.

Closely related to the issue of flux containment is leakage inductance. We've already seen the detrimental effects of leakage inductance on voltage regulation with SPICE simulations early in this chapter. Because leakage inductance is equivalent to an inductance connected in series with the transformer's winding, it manifests itself as a series impedance with the load. Thus, the more current drawn by the load, the less voltage available at the secondary winding terminals. Usually, good voltage regulation is desired in transformer design, but there are exceptional applications.

As was stated before, discharge lighting circuits require a step-up transformer with"loose" (poor) voltage regulation to ensure reduced voltage after the establishment of an arc through the lamp. One way to meet this design criterion is to engineer the transformer with flux leakage paths for magnetic flux to bypass the secondary winding(s). The resulting leakage flux will produce leakage inductance, which will in turn produce the poor regulation needed for discharge lighting.

CORE SATURATION

Transformers are also constrained in their performance by the magnetic flux limitations of the core. For ferromagnetic core transformers, we must be mindful of the saturation limits of the core. Remember that ferromagnetic materials cannot support infinite magnetic flux densities: they tend to"saturate" at a certain level (dictated by the material and core dimensions), meaning that further increases in magnetic field force (mmf) do not result in proportional increases in magnetic field flux (Φ). When a transformer's primary winding is overloaded from excessive applied voltage, the core flux may reach saturation levels during peak moments of the AC sinewave cycle. If this happens, the voltage induced in the secondary winding will no longer match the wave-shape as the voltage powering the primary coil. In other words, the overloaded transformer will distort the waveshape from primary to secondary windings, creating harmonics in the secondary winding's output. As we discussed before, harmonic content in AC power systems typically causes problems.

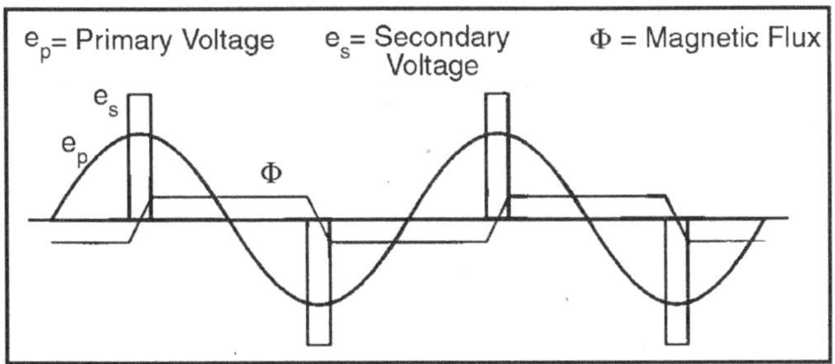

Fig. Voltage and Flux Waveforms for a Peaking Transformer.

Special transformers known as peaking transformers exploit this principle to produce brief voltage pulses near the peaks of the source voltage waveform. The core is designed to saturate quickly and sharply, at voltage levels well below peak. This results in a severely cropped sine-wave flux waveform, and secondary voltage pulses only when the flux is changing (below saturation levels):

Another cause of abnormal transformer core saturation is operation at frequencies lower than normal. For example, if a power transformer designed to operate at 60 Hz is forced to operate at 50 Hz instead, the flux must reach greater peak levels than before in order to produce the same opposing voltage needed to balance against the source voltage. This is true even if the source voltage is the same as before.

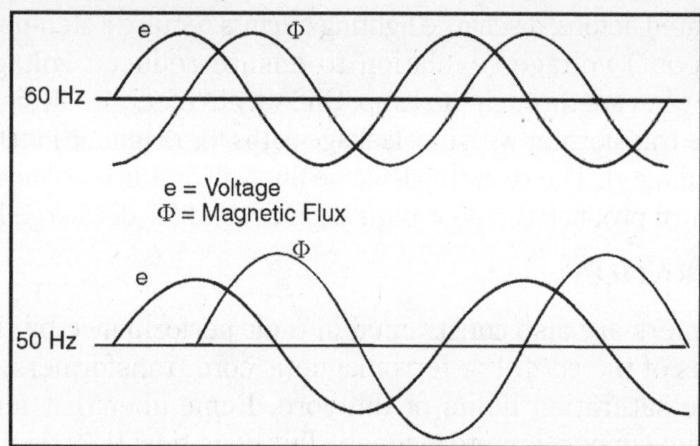

Fig. Magnetic Flux is Higher in a Transformer core Driven by 50 Hz as Compared to 60 Hz for the same Voltage.

Since instantaneous winding voltage is proportional to the instantaneous magnetic flux's rate of change in a transformer, a voltage waveform reaching the same peak value, but taking a longer amount of time to complete each half-cycle, demands that the flux maintain the same rate of change as before, but for longer periods of time. Thus, if the flux has to climb at the same rate as before, but for longer periods of time, it will climb to a greater peak value. Mathematically, this is another example of calculus in action.

Because the voltage is proportional to the flux's rate-of-change, we say that the voltage waveform is the derivative of the flux waveform,"derivative" being that calculus operation defining one mathematical function (waveform) in terms of the rate-of-change of another. If we take the opposite perspective, though, and relate the original waveform to its derivative, we may call the original waveform the integral of the derivative waveform. In this case, the voltage waveform is the derivative of the flux waveform, and the flux waveform is the integral of the voltage waveform. The integral of any mathematical function is proportional to the area accumulated underneath the curve of that function. Since each half-cycle of the 50 Hz waveform accumulates more area between it and the zero line of the graph than the 60 Hz waveform will -- and we know that the magnetic flux is the integral of the voltage -- the flux will attain higher values.

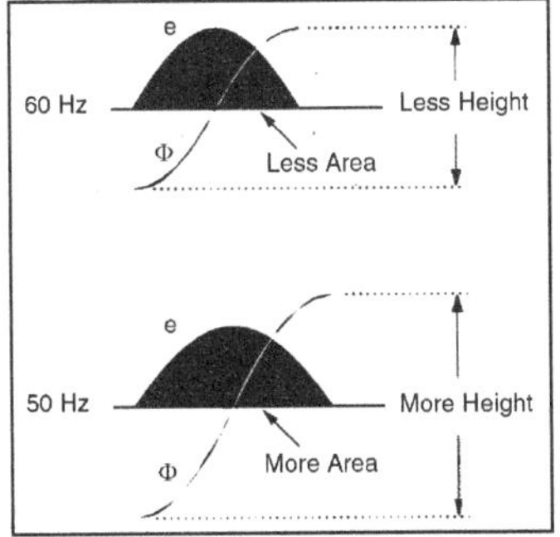

Fig. Flux Changing at the same rate Rises to a Higher level at 50 Hz than at 60 Hz.

Yet another cause of transformer saturation is the presence of DC current in the primary winding. Any amount of DC voltage dropped across the primary winding of a transformer will cause additional magnetic flux in the core.This additional flux"bias" or"offset" will push the alternating flux waveform closer to saturation in one half-cycle than the other.

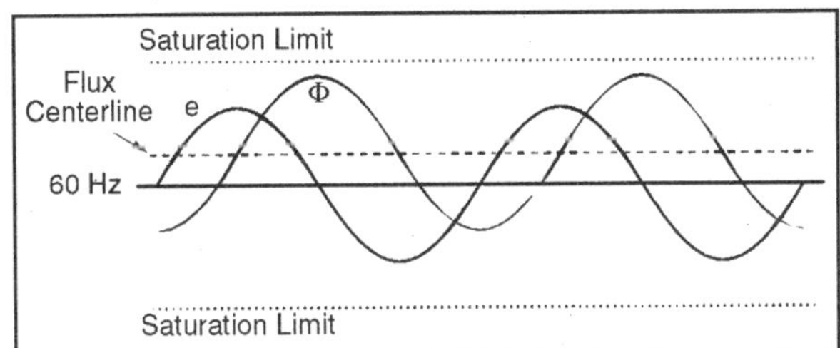

Fig. DC in Primary, Shifts the Waveform peaksTowards the Upper Saturation limit.

For most transformers, core saturation is a very undesirable effect, and it is avoided through good design: engineering the windings and core so that magnetic flux densities remain well below the saturation levels.

INRUSH CURRENT

When a transformer is initially connected to a source of AC voltage,

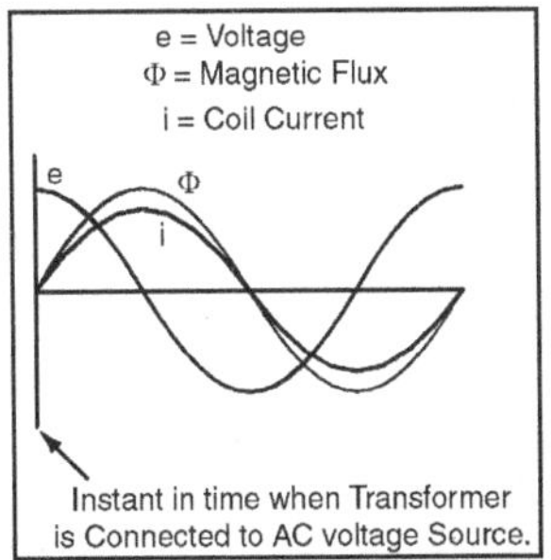

Fig.*Connecting Transformer to Line at AC Volt Peak*: Flux Increases Rapidly from Zero, same as Steady-state Operation.

Alternatively, let us consider what happens if the transformer's connection to the AC voltage source occurs at the exact moment in time when the instantaneous voltage is at zero. During continuous operation (when the transformer has been powered for quite some time), this is the point in time where both flux and winding current are at their negative peaks, experiencing zero rate-of-change ($d\Phi/dt = 0$ and $di/dt = 0$). As the voltage builds to its positive peak, the flux and current waveforms build to their maximum positive rates-of-change, and on upward to their positive peaks as the voltage descends to a level of zero:

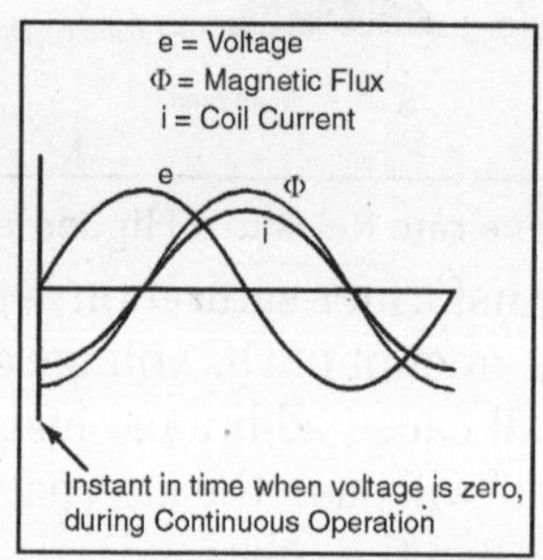

Fig. Starting at e=0 V is not the same as Running Continuously in. These Expected Waveforms are Incorrect- F and i should start at zero.

A significant difference exists, however, between continuous-mode operation and the sudden starting condition assumed in this scenario: during continuous operation, the flux and current levels were at their negative peaks when voltage was at its zero point; in a transformer that has been sitting idle, however, both magnetic flux and winding current should start at zero.

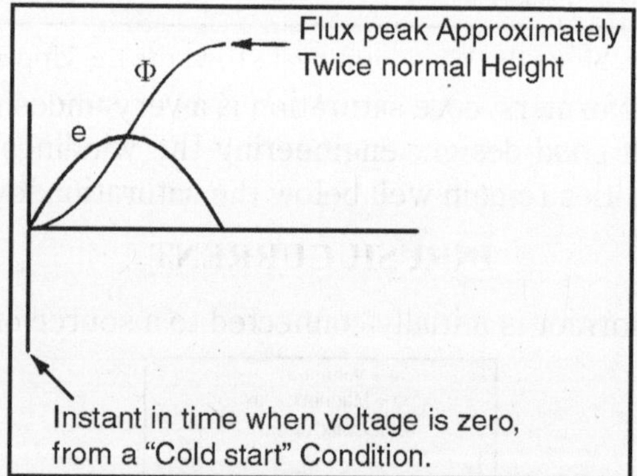

Fig. Starting at e=0 V, F starts at Initial Condition F=0, Increasing to Twice the Normal value, Assuming it doesn't Saturate the core.

When the magnetic flux increases in response to a rising voltage, it will increase from zero upward, not from a previously negative (magnetized) condition as we would normally have in a transformer that's been powered for awhile. Thus, in a transformer that's just"starting," the flux will reach approximately twice its normal peak magnitude as it"integrates" the area under the voltage waveform's first half-cycle:

In an ideal transformer, the magnetizing current would rise to approximately twice its normal peak value as well, generating the necessary mmf to create this higher-than-normal flux.

However, most transformers aren't designed with enough of a margin between normal flux peaks and the saturation limits to avoid saturating in a condition like this, and so the core will almost certainly saturate during this first half-cycle of voltage.

During saturation, disproportionate amounts of mmf are needed to generate magnetic flux. This means that winding current, which creates the mmf to cause flux in the core, will disproportionately rise to a value easily exceeding twice its normal peak:

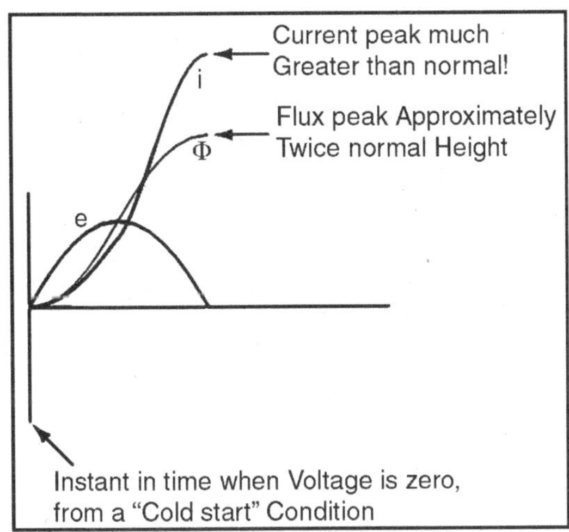

Fig. Starting at e=0 V, Current also Increases to Twice the Normal value for an Unsaturated core, or Considerably Higher in the (Designed for) case of Saturation.

This is the mechanism causing inrush current in a transformer's primary winding when connected to an AC voltage source. As you can see, the magnitude of the inrush current strongly depends on the exact time that electrical connection to the source is made.

If the transformer happens to have some residual magnetism in its core at the moment of connection to the source, the inrush could be even more severe. Because of this, transformer overcurrent protection devices are usually of the"slow-acting" variety, so as to tolerate current surges such as this without opening the circuit.

HEAT AND NOISE

In addition to unwanted electrical effects, transformers may also exhibit undesirable physical effects, the most notable being the production of heat and noise. Noise is primarily a nuisance effect, but heat is a potentially serious problem because winding insulation will be damaged if allowed to overheat.

Heating may be minimized by good design, ensuring that the core does not approach saturation levels, that eddy currents are minimized, and that the windings are not overloaded or operated too close to maximum ampacity.

Large power transformers have their core and windings submerged in an oil bath to transfer heat and muffle noise, and also to displace moisture which would otherwise compromise the integrity of the winding insulation. Heat-dissipating"radiator" tubes on the outside of the transformer case provide a convective oil flow path to transfer heat from the transformer's core to ambient air: (Figure below)

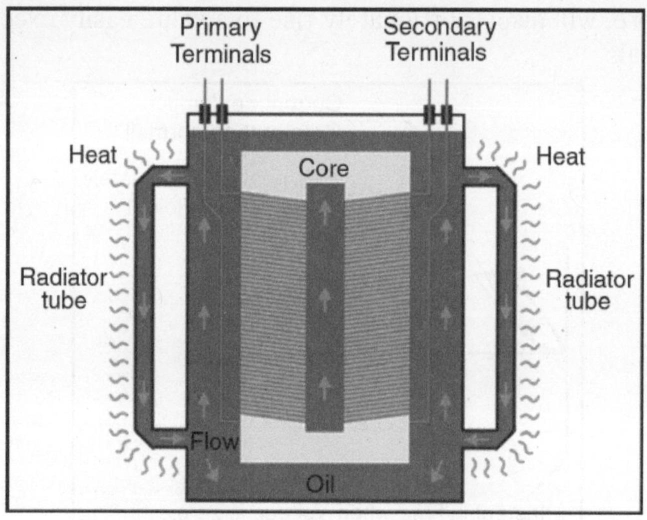

Fig. Large Power Transformers are Submerged in Heat Dissipating Insulating oil.

Oil-less, or"dry," transformers are often rated in terms of maximum operating temperature"rise" (temperature increase beyond ambient) according to a letter-class system. *A, B, F, or H. These letter codes are arranged in order of lowest heat tolerance to highes*t:

- *Class A*: No more than 55° Celsius winding temperature rise, at 40° Celsius (maximum) ambient air temperature.
- *Class B*: No more than 80° Celsius winding temperature rise, at 40° Celsius (maximum)ambient air temperature.
- *Class F*: No more than 115° Celsius winding temperature rise, at 40° Celsius (maximum)ambient air temperature.
- *Class H*: No more than 150° Celsius winding temperature rise, at 40° Celsius (maximum)ambient air temperature.

Audible noise is an effect primarily originating from the phenomenon of magnetostriction: the slight change of length exhibited by a ferromagnetic object when magnetized. The familiar"hum" heard around large power transformers is the sound of the iron core expanding and contracting at 120 Hz (twice the system frequency, which is 60 Hz in the United States) -- one cycle of core contraction and expansion for every peak of the magnetic flux waveform -- plus noise created by mechanical forces between primary and secondary windings. Again, maintaining low magnetic flux levels in the core is the key to minimizing this effect, which explains why ferroresonant transformers -- which must operate in saturation for a large portion of the current waveform -- operate both hot and noisy.

Another noise-producing phenomenon in power transformers is the physical reaction force between primary and secondary windings when heavily loaded. If the secondary winding is open-circuited, there will be no current through it, and consequently no magneto-motive force (mmf) produced by it. However, when the secondary is"loaded" (current supplied to a load), the winding generates an mmf, which becomes counteracted by a"reflected" mmf in the primary winding to prevent core flux levels from changing.

These opposing mmf's generated between primary and secondary windings as a result of secondary (load) current produce a repulsive, physical force between the windings which will tend to make them vibrate. Transformer designers have to consider these physical forces in the construction of the winding coils, to ensure there is adequate mechanical support to handle the stresses. Under heavy load (high current) conditions, though, these stresses may be great enough to cause audible noise to emanate from the transformer.

This ensures that the relationship between mmf and ? is more linear throughout the flux cycle, which is good because it makes for less distortion in the magnetization current waveform.

4

Generation of Power System

Power system engineering forms a vast and major portion of electrical engineering studies. It is mainly concerned with the production of electrical power and its transmission from the sending end to the receiving end as per consumer requirements, incurring minimum amount of losses. The power at the consumer end is often subjected to changes due to the variation of load or due to disturbances induced within the length of transmission line. For this reason the term power system stability is of utmost importance in this field, and is used to define the ability of the of the system to bring back its operation to steady state condition within minimum possible time after having undergone some sort of transience or disturbance in the line.

Ever since, the 20th century, till the recent times all major power generating stations over the globe has mainly relied on A.C. distribution system as the most effective and economical option for the transmission of electrical power. Even the most effective way to produce bulk amount of power has been with the evolution of A.C. machine (*i.e.,* alternator or synchronous generator). In the power plants, several synchronous generators with different voltage ratings are connected to the bus terminals having the same frequency and phase sequence as the generators, while the consumer ends are feeded directly from those bus terminals.

And therefore for stable operation it is important for the bus to be well synchronized with the generators over the entire duration of transmission, and for this reason the power system stability is also referred to as synchronous stability and is defined as the ability of the system to return to synchronism after having undergone some disturbance due to switching on and off of load or due to line transience.

To understand stability well another factor that is to be taken into consideration is the stability limit of the system. The stability limit defines the maximum power permissible to flow through a particular point or a part of the system during which it is subjected to line disturbances or faulty flow of power. Having understood these terminologies related to power system stability let us now look into the different types of stability.

The synchronous stability of a power system can be of several types depending upon the nature of disturbance, and for the purpose of successful analysis it can be classified into the following 3 types as shown below:

1. Steady state stability.
2. Transient stability.
3. Dynamic stability.

STEADY STATE STABILITY OF A POWER SYSTEM

The steady state stability of a power system is defined as the ability of the system to bring itself back to its stable configuration following a small disturbance in the network (like normal load fluctuation or action of automatic voltage regulator).

It can only be considered only during a very gradual and infinitesimally small power change. In case the power flow through the circuit exceeds the maximum power permissible, then there are chances that a particular machine or a group of machines will cease to operate in synchronism, and result in yet more disturbances. In such a situation, the steady state limit of the system is said to have reached. Or in other words the steady state stability limit of a system refers to the maximum amount of power that is permissible through the system without loss of its steady state stability.

TRANSIENT STABILITY OF A POWER SYSTEM

Transient stability of a power systemrefers to the ability of the system to reach a stable condition following a large disturbance in the network condition. In all cases related to large changes in the system like sudden application or removal of load, switching operations, line faults or loss due to excitation the transient stability of the system comes into play. It infact deals in the ability of the system to retain synchronism following a disturbance sustaining for a reasonably long period of time. And the maximum power that is permissible to flow through the network without loss of stability following a sustained period of disturbance is referred to as the transient stability of the system. Going beyond that maximum permissible value for power flow, the system would temporarily be rendered as unstable.

DYNAMIC STABILITY OF A POWER SYSTEM

Dynamic stability of a system denotes the artificial stability given to an inherently unstable system by automatic controlled means. It is generally concerned to small disturbances lasting for about 10 to 30 seconds.

PRACTICE OF POWER SYSTEMS

Despite their common components, power systems vary widely both with respect to their design and how they operate.

RESIDENTIAL POWER SYSTEMS

Residential dwellings almost always take supply from the low voltage distribution lines or cables that run past the dwelling. These operate at voltages of between 110 and 260 volts (phase-to-earth) depending upon national standards. A few decades ago small dwellings would be fed a single phase using a dedicated two-core service cable (one core for the active phase and one core for the neutral return). The active line would then be run through a main isolating switch in the fuse box and then split into one or more circuits to feed lighting and appliances inside the house. By convention, the lighting and appliance circuits are kept separate so the failure of an appliance does not leave the dwelling's occupants in the dark. All circuits would be fused with an appropriate fuse based upon the wire size used for that circuit. Circuits would have both an active and neutral wire with both the lighting and power sockets being connected in parallel. Sockets would also be provided with a protective earth. This would be made available to appliances to connect to any metallic casing. If this casing were to become live, the theory is the connection to earth would cause an RCD or fuse to trip - thus preventing the future electrocution of an occupant handling the appliance. Earthing systems vary between regions, but in countries such as the United Kingdom and Australia both the protective earth and neutral line would be earthed together near the fuse box before the main isolating switch and the neutral earthed once again back at the distribution transformer.

There have been a number of minor changes over the year to practice of residential wiring. Some of the most significant ways modern residential power systems tend to vary from older ones include:

- For convenience, miniature circuit breakers are now almost always used in the fuse box instead of fuses as these can easily be reset by occupants.
- For safety reasons, RCDs are now installed on appliance circuits and, increasingly, even on lighting circuits.
- Dwellings are typically connected to all three-phases of the distribution system with the phases being arbitrarily allocated to the house's single-phase circuits.
- Whereas air conditioners of the past might have been fed from a dedicated circuit attached to a single phase, centralised air conditioners that require three-phase power are now becoming common.
- Protective earths are now run with lighting circuits to allow for metallic lamp holders to be earthed.
- Increasingly residential power systems are incorporating microgenerators, most notably, photovoltaic cells.

COMMERCIAL POWER SYSTEMS

Commercial power systems such as shopping centers or high-rise buildings are larger in scale than residential systems. Electrical designs for larger commercial systems are usually studied for load flow, short-circuit fault levels, and voltage drop for steady-state loads and during starting of large motors. The objectives of the studies are to assure proper equipment and conductor sizing, and to coordinate protective devices so that minimal disruption is cause when a fault is cleared. Large commercial installations will have an orderly system of sub-panels, separate from the main distribution board to allow for better system protection and more efficient electrical installation. Typically one of the largest appliances connected to a commercial power system is the HVAC unit, and ensuring this unit is adequately supplied is an important consideration in commercial power systems. Regulations for commercial establishments place other requirements on commercial systems that are not placed on residential systems. For example, in Australia, commercial systems must comply with AS 2293, the standard for emergency lighting, which requires emergency lighting be maintained for at least 90 minutes in the event of loss of mains supply. In the United States, the National Electrical Code requires commercial systems to be built with at least one 20A sign outlet in order to light outdoor signage. Building code regulations may place special requirements on the electrical system for emergency lighting, evacuation, emergency power, smoke control and fire protection.

EXPLANATION OF POWER SYSTEM

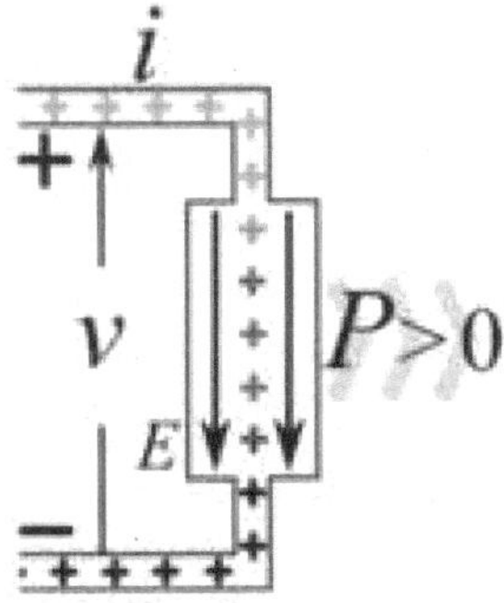

Fig. Animation showing electric load

Electric power is transformed to other forms of power when electric charges move through an electric potential (voltage) difference, which occurs inelectrical components in electric circuits. From the standpoint of electric power, components in an electric circuit can be divided into two categories:

- Passive devices or loads: When electric charges move through a potential difference from a high voltage to a low voltage, that is conventional current (positive charge) moves from the positive terminal to the negative, the potential energy of the charges is converted to kinetic energy, which performs work on the device. Devices in which this occurs are called *passive* devices or *loads*; they consume electric power from the circuit, converting it to other forms such as mechanical work, heat, light, etc. Examples are electrical appliances, such as light bulbs, electric motors, andelectric heaters. In alternating current (AC) circuits the direction of the current and voltage periodically reverses, but the instantaneous current is always moving from the high potential to the low potential side.

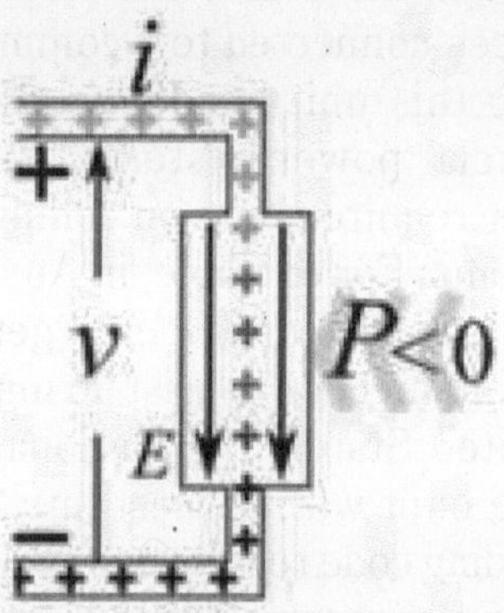

Fig. Animation showing power source

- Active devices or power sources: If the charges are forced by an outside force to move through the device in the direction from a lower electric potential to a higher, so positive charge moves from the negative to the positive terminal, work is being done *on* the charges, so energy is being converted to electric potential energy from some other type of energy, such as mechanical energy or chemical energy. Devices in which this occurs are called *active* devices or *power sources*; sources of electric current, such as electric generators and batteries.

Some devices can be either a source or a load, depending on the voltage or current through them.

For example, a rechargeable battery acts as a source when it provides power to a circuit, but as a load when it is connected to a battery charger and is being recharged.

PASSIVE SIGN CONVENTION

Since electric power can flow in either direction, either into or out of a component, a convention is needed for which direction represents positive power flow. Electric power flowing *out* of a circuit *into* a component is arbitrarily defined to have a positive sign, while power flowing *into* a circuit from a component is defined to have a negative sign. Thus passive components have positive power consumption. This is called the *passive sign convention.*

RESISTIVE CIRCUITS

In the case of resistive (Ohmic, or linear) loads, Joule's law can be combined with Ohm's law ($V = I{\cdot}R$) to produce alternative expressions for the dissipated power:

$$P = I^2R = \frac{V^2}{R},$$

where R is the electrical resistance.

ALTERNATING CURRENT

In alternating current circuits, energy storage elements such as inductance and capacitance may result in periodic reversals of the direction of energy flow. The portion of power flow that, averaged over a complete cycle of the AC waveform, results in net transfer of energy in one direction is known as real power (also referred to as active power). That portion of power flow due to stored energy, that returns to the source in each cycle, is known as reactive power. The real power P in watts consumed by a device is given by

$$P = \frac{1}{2}V_pI_p\cos\theta = V_{rms}I_{rms}\cos\theta$$

where

V_p is the peak voltage in volts

I_p is the peak current in amperes

V_{rms} is the root-mean-square voltage in volts

I_{rms} is the root-mean-square current in amperes

è is the phase angle between the current and voltage sine waves

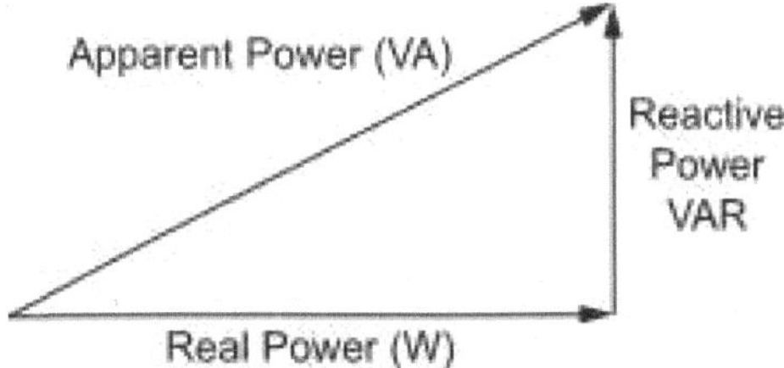

Power triangle: The components ofAC power

The relationship between real power, reactive power and apparent power can be expressed by representing the quantities as vectors.

Real power is represented as a horizontal vector and reactive power is represented as a vertical vector. The apparent power vector is the hypotenuse of a right triangle formed by connecting the real and reactive power vectors. This representation is often called the *power triangle*. Using the Pythagorean

Theorem, the relationship among real, reactive and apparent power is:

$$(\text{apparent power})^2 = (\text{real power})^2 + (\text{reactive power})^2$$

Real and reactive powers can also be calculated directly from the apparent power, when the current and voltage are both sinusoids with a known phase angle è between them:

$$(\text{real power}) = (\text{apparent power})\cos\theta$$

$$(\text{reactive power}) = (\text{apparent power})\sin\theta$$

The ratio of real power to apparent power is called power factor and is a number always between 0 and 1. Where the currents and voltages have non-sinusoidal forms, power factor is generalized to include the effects of distortion

ELECTROMAGNETIC FIELDS

Electrical energy flows wherever electric and magnetic fields exist together and fluctuate in the same place. The simplest example of this is in electrical circuits, as the preceding.

In the general case, however, the simple equation $P = IV$ must be replaced by a more complex calculation, the integral of the cross-product of the electrical and magnetic field vectors over a specified area, thus:

$$P = \int_S (\mathbf{E} \times \mathbf{H}) \cdot \mathbf{dA}.$$

The result is a scalar since it is the *surface integral* of the *Poynting vector.*

COMPONENTS OF POWER SYSTEMS

SUPPLIES

The majority of the world's power still comes from coal-fired power stations like this.

All power systems have one or more sources of power. For some power systems, the source of power is external to the system but for others it is part of the system itself - it is these internal power sources that are discussed in the remainder. Direct current power can be supplied by batteries, fuel cells or photovoltaic cells. Alternating current power is typically supplied by a rotor that spins in a magnetic field in a device known as a turbo generator. There have been a wide range of techniques used to spin a turbine's rotor, from steam heated using fossil fuel (including coal, gas and oil) or nuclear energy, falling water (hydroelectric power) and wind (wind power).

The speed at which the rotor spins in combination with the number of generator poles determines the frequency of the alternating current produced by the generator. All generators on a single synchronous system, for example the national grid, rotate at sub-multiples of the same speed and so generate electrical current at the same frequency. If the load on the system increases, the generators will require more torque to spin at that speed and, in a typical power station, more steam must be supplied to the turbines driving them. Thus the steam used and the fuel expended are directly dependent on the quantity of electrical energy supplied. An exception exists for generators incorporating power electronics such as gearless wind turbines or linked to a grid through an asynchronous tie such as a HVDC link — these can operate at frequencies independent of the power system frequency.

Depending on how the poles are fed, alternating current generators can produce a variable number of phases of power. A higher number of phases leads to more efficient power system operation but also increases the infrastructure requirements of the system.

Electricity grid systems connect multiple generators and loads operating at the same frequency and number of phases, the commonest being three-phase at 50 or 60 Hz. However there are other considerations. These range from the obvious: How much power should the generator be able to supply? What is an acceptable length of time for starting the generator (some generators can take hours to start)? Is the availability of the power source acceptable (some renewables are only available when the sun is shining or the wind is blowing)? To the more technical: How should the generator start (some turbines act like a motor to bring themselves up to speed in which case they need an appropriate starting circuit)? What is the mechanical speed of operation for the turbine and consequently what are the number of poles required? What type of generator is suitable (synchronous or asynchronous) and what type of rotor (squirrel-cage rotor, wound rotor, salient pole rotor or cylindrical rotor)?

LOADS

A toaster is great example of a single-phase load that might appear in a residence. Toasters typically draw 2 to 10 amps at 110 to 260 volts consuming around 600 to 1200 watts of power Power systems deliver energy to loads that perform a function. These loads range from household appliances to industrial machinery. Most loads expect a certain voltage and, for alternating current devices, a certain frequency and number of phases. The appliances found in your home, for example, will typically be single-phase operating at 50 or 60 Hz with a voltage between 110 and 260 volts (depending on national standards). An exception exists for centralized air conditioning systems as these are now typically three-phase because this allows them to operate more efficiently. All devices in your house will also have a wattage, this specifies the amount of power the device consumes. At any one time, the net amount of power consumed by the loads on a power system must equal the net amount of power produced by the supplies less the power lost in transmission.

Making sure that the voltage, frequency and amount of power supplied to the loads is in line with expectations is one of the great challenges of power system engineering. However it is not the only challenge, in addition to the power used by a load to do useful work (termed real power) many alternating current devices also use an additional amount of power because they cause the alternating voltage and alternating current to become slightly out-of-sync (termed reactive power).

The reactive power like the real power must balance (that is the reactive power produced on a system must equal the reactive power consumed) and can be supplied from the generators, however it is often more economical to supply such power from capacitors.

A final consideration with loads is to do with power quality. In addition to sustained overvoltages and undervoltages (voltage regulation issues) as well as sustained deviations from the system frequency (frequency regulation issues), power system loads can be adversely affected by a range of temporal issues.

These include voltage sags, dips and swells, transient overvoltages, flicker, high frequency noise, phase imbalance and poor power factor. Power quality issues occur when the power supply to a load deviates from the ideal: For an AC supply, the ideal is the current and voltage in-sync fluctuating as a perfect sine wave at a prescribed frequency with the voltage at a prescribed amplitude. For DC supply, the ideal is the voltage not varying from a prescribed level. Power quality issues can be especially important when it comes to specialist industrial machinery or hospital equipment.

CONDUCTORS

Conductors carry power from the generators to the load. In a grid, conductors may be classified as belonging to the transmission system, which carries large amounts of power at high voltages (typically more than 69 kV) from the generating centres to the load centres, or the distribution system, which feeds smaller amounts of power at lower voltages (typically less than 69 kV) from the load centres to nearby homes and industry.

Choice of conductors is based upon considerations such as cost, transmission losses and other desirable characteristics of the metal like tensile strength. Copper, with lower resistivity than aluminium, was the conductor of choice for most power systems. However, aluminum has lower cost for the same current carrying capacity and is the primary metal used for transmission line conductors. Overhead line conductors may be reinforced with steel or aluminum alloys.

Conductors in exterior power systems may be placed overhead or underground. Overhead conductors are usually air insulated and supported on porcelain, glass or polymer insulators. Cables used for underground transmission or building wiring are insulated with cross-linked polyethylene or other flexible insulation. Large conductors are stranded for ease of handling; small conductors used for building wiring are often solid, especially in light commercial or residential construction.

Conductors are typically rated for the maximum current that they can carry at a given temperature rise over ambient conditions. As current flow increases through a conductor it heats up. For insulated conductors, the rating is determined by the insulation. For overhead conductors, the rating is determined by the point at which the sag of the conductors would become unacceptable.

CAPACITORS AND REACTORS

The majority of the load in a typical AC power system is inductive; the current lags behind the voltage. Since the voltage and current are out-of-phase, this leads to the emergence of an "imaginary" form of power known as reactive power. Reactive power does no measurable work but is transmitted back and forth between the reactive power source and load every cycle. This reactive power can be provided by the generators themselves, through the adjustment of generator excitation, but it is often cheaper to provide it through capacitors, hence capacitors are often placed near inductive loads to reduce current demand on the power system (*i.e.*, increase the power factor), which may never exceed 1.0, and which represents a purely resistive load. Power factor correction may be applied at a central substation, through the use of so-called "synchronous condensers" (synchronous machines which act as condensers which are variable in VAR value, through the adjustment of machine excitation) or adjacent to large loads, through the use of so-called "static condensers" (condensers which are fixed in VAR value). Reactors consume reactive power and are used to regulate voltage on long transmission lines. In light load conditions, where the loading on transmission lines is well below thesurge impedance loading, the efficiency of the power system may actually be improved by switching in reactors. Reactors installed in series in a power system also limit rushes of current flow, small reactors are therefore almost always installed in series with capacitors to limit the current rush associated with switching in a capacitor. Series reactors can also be used to limit fault currents.

Capacitors and reactors are switched by circuit breakers, which results in moderately large steps in reactive power. A solution comes in the form of static VAR compensators andstatic synchronous compensators. Briefly, static VAR compensators work by switching in capacitors using thyristors as opposed to circuit breakers allowing capacitors to be switched-in and switched-out within a single cycle. This provides a far more refined response than circuit breaker switched capacitors. Static synchronous compensators take a step further by achieving reactive power adjustments using only power electronics.

POWER ELECTRONICS

Power electronics are semi-conductor based devices that are able to switch quantities of power ranging from a few hundred watts to several hundred megawatts. Despite their relatively simple function, their speed of operation (typically in the order of nanoseconds) means they are capable of a wide range of tasks that would be difficult or impossible with conventional technology. The classic function of power electronics is rectification, or the conversion of AC-to-DC power, power electronics are therefore found in almost every digital device that is supplied from an AC source either as an adapter that plugs into the wall or as component internal to the device. High-powered power electronics can also be used to convert AC power to DC power for long distance transmission in a system known as HVDC. HVDC is used because it proves to be more economical than similar high voltage AC systems for very long distances (hundreds to thousands of kilometres). HVDC is also desirable for interconnects because it allows frequency independence thus improving system stability. Power electronics are also essential for any power source that is required to produce an AC output but that by its nature produces a DC output. They are therefore used by many photovoltaic installations both industrial and residential.

Power electronics also feature in a wide range of more exotic uses. They are at the heart of all modern electric and hybrid vehicles - where they are used for both motor control and as part of the brushless DC motor. Power electronics are also found in practically all modern petrol-powered vehicles, this is because the power provided by the car's batteries alone is insufficient to provide ignition, air-conditioning, internal lighting, radio and dashboard displays for the life of the car. So the batteries must be recharged while driving using DC power from the engine - a feat that is typically accomplished using power electronics. Whereas conventional technology would be unsuitable for a modern electric car, commutators can and have been used in petrol-powered cars, the switch to alternators in combination with power electronics has occurred because of the improved durability of brushless machinery.

Some electric railway systems also use DC power and thus make use of power electronics to feed grid power to the locomotives and often for speed control of the locomotive's motor. In the middle twentieth century, rectifier locomotives were popular, these used power electronics to convert AC power from the railway network for use by a DC motor.Today most electric locomotives are supplied with AC power and run using AC motors, but still use power electronics to provide suitable motor control. The use of power electronics to assist with motor control and with starter circuits cannot be underestimated and, in addition to rectification, is responsible for power electronics appearing in a wide range of industrial machinery. Power electronics even appear in modern residential air conditioners.

Power electronics are also at the heart of the variable speed wind turbine. Conventional wind turbines require significant engineering to ensure they operate at some ratio of the system frequency, however by using power electronics this requirement can be eliminated leading to quieter, more flexible and (at the moment) more costly wind turbines. A final example of one of the more exotic uses of power electronics comes the fast-switching times of power electronics were used to provide more refined reactive compensation to the power system.

PROTECTIVE DEVICES

Power systems contain protective devices to prevent injury or damage during failures. The quintessential protective device is the fuse. When the current through a fuse exceeds a certain threshold, the fuse element melts, producing an arc across the resulting gap that is then extinguished, interrupting the circuit. Given that fuses can be built as the weak point of a system, fuses are ideal for protecting circuitry from damage. Fuses however have two problems: First, after they have functioned, fuses must be replaced as they cannot be reset. This can prove inconvenient if the fuse is at a remote site or a spare fuse is not on hand. And second, fuses are typically inadequate as the sole safety device in most power systems as they allow current flows well in excess of that that would prove lethal to a human or animal.

The first problem is resolved by the use of circuit breakers - devices that can be reset after they have broken current flow. In modern systems that use less than about 10 kW, miniature circuit breakers are typically used. These devices combine the mechanism that initiates the trip (by sensing excess current) as well as the mechanism that breaks the current flow in a single unit. Some miniature circuit breakers operate solely on the basis of electromagnetism. In these miniature circuit breakers, the current is run through a solenoid, and, in the event of excess current flow, the magnetic pull of the solenoid is sufficient to force open the circuit breaker's contacts (often indirectly through a tripping mechanism). A better design however arises by inserting a bimetallic strip before the solenoid - this means that instead of always producing a magnetic force, the solenoid only produces a magnetic force when the current is strong enough to deform the bimetallic strip and complete the solenoid's circuit.

In higher powered applications, the protective relays that detect a fault and initiate a trip are separate from the circuit breaker. Early relays worked based upon electromagnetic principles similar to those mentioned in the previous paragraph, modern relays are application-specific computers that determine whether to trip based upon readings from the power system. Different relays will initiate trips depending upon different protection schemes. For example, an overcurrent relay might initiate a trip if the current on any phase exceeds a certain threshold whereas a set of differential relays might initiate a trip if the sum of currents between them indicates there may be current leaking to earth. The circuit breakers in higher powered applications are different too. Air is typically no longer sufficient to quench the arc that forms when the contacts are forced open so a variety of techniques are used. One of the most popular techniques is to keep the chamber enclosing the contacts flooded with sulfur hexafluoride (SF_6) - a non-toxic gas that has sound arc-quenching properties. Other techniques are discussed in the reference.

The second problem, the inadequacy of fuses to act as the sole safety device in most power systems, is probably best resolved by the use of residual current devices (RCDs). In any properly functioning electrical appliance the current flowing into the appliance on the active line should equal the current flowing out of the appliance on the neutral line. A residual current device works by monitoring the active and neutral lines and tripping the active line if it notices a difference. Residual current devices require a separate neutral line for each phase and to be able to trip within a time frame before harm occurs. This is typically not a problem in most residential applications where standard wiring provides an active and neutral line for each appliance (that's why your power plugs always have at least two tongs) and the voltages are relatively low however these issues do limit the effectiveness of RCDs in other applications such as industry. Even with the installation of an RCD, exposure to electricity can still prove lethal.

SCADA SYSTEMS

In large electric power systems, Supervisory Control And Data Acquisition

(SCADA) is used for tasks such as switching on generators, controlling generator output and switching in or out system elements for maintenance. The first supervisory control systems implemented consisted of a panel of lamps and switches at a central console near the controlled plant. The lamps provided feedback on the state of plant (the data acquisition function) and the switches allowed adjustments to the plant to be made (the supervisory control function). Today, SCADA systems are much more sophisticated and, due to advances in communication systems, the consoles controlling the plant no longer need to be near the plant itself. Instead it is now common for plant to be controlled from a with equipment similar to (if not identical to) a desktop computer. The ability to control such plant through computers has increased the need for security and already there have been reports of cyber-attacks on such systems causing significant disruptions to power systems.

GENERATOR LOAD CONTROL LOOPS

This chapter represents the effect of loads on the voltage and frequency. The system frequency deviation affects the changes of real power and reactive power is affected by the voltage magnitude and frequency.

CONTROL LOOP

Changes in real power affect mainly the system frequency, while reactive power is less sensitive to changes in frequency and is mainly depended on changes in voltage magnitude. Thus the real and reactive powers are controlled separately. The load frequency control (LFC) loop controls the real power frequency, and automatic voltage regulator (AVR) loop regulator the reactive power and voltage magnitude .

Basic Generator Control Loops

An isolated and interconnected power system, Load frequency Control (LFC) and Automatic Voltage regulator (AVR) equipments are installed for each generator.

The schematic diagram of the Load Frequency Control loops and the Automatic Voltage regulator (AVR) control loops of a synchronous generator. The controllers are set frequency operating condition and take care of small changes in load demand to the frequency and voltage magnitude within the specified limits. Small changes in real power are mainly dependent on changes in the rotor angle d and thus the frequency. Reactive power is mainly depends on the voltage magnitude (*i.e.,* on the generator excitation).the excitation system time constant is much smaller than the prime over constant and its transient decay is much faster and does not affect the LFC dynamic. Thus, the cross-coupling between the LFC loop and the AVR loop is negligible, and the load frequency and excitation voltage control are analyzed independently.

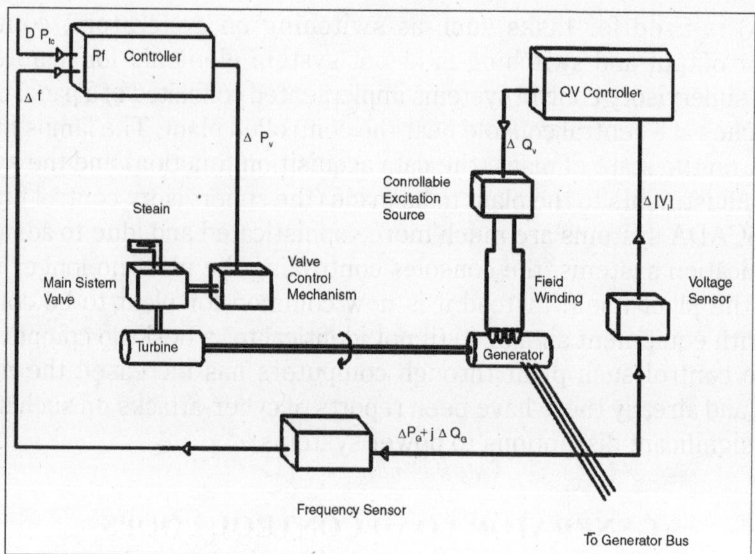

Fig. Load Frequency Control Loops and the Automatic Voltage Regulator (AVR) Control Loops of a Synchronous Generator.

MEGAWATT-FREQUENCY, OR P-F CONTROL LOOPS

The objective of this loop this is to exert control of frequency and simultaneously of the real power exchange via outgoing lines. Frequency sensor senses the frequency 'error' ?f and the increments in the real tie line powers, which will indirectly provide information about the incremental static error, ?d. These sensor signals are amplified, mixed and transformed into a real power command signal in ?PV which is sent to the prime mover to the call for an increment ithen change the state increments sensed.

MEGAVAR-VOLTAGE, OR Q-V CONTROL LOOPS

The objective of this loop this is to exert control of the voltage state |Vi|. The voltage error Δ|Vi| is sensed and this signal is transformed into a reactive power command signal ΔQV, which is fed to excitation source. The result is a change in the rotor current, and thus in the generator EMF, which finally adds up to an increment change in the reactive generator ΔQG.

DYNAMIC INTERACTION BETWEEN P-F AND Q-V CONTROL LOOPS

In static sense, and for small deviations, there is little interaction between the p-f and Q-V control loops .

During dynamic perturbations we encounter considerable coupling between the two control channels, for two different responses.

- As the voltage magnitude fluctuates at a bus, the real load of that bus will likewise change as a result of the voltage load characteristics ($\partial P_D/\partial|V|$).

- As the voltage magnitude fluctuates at a bus, the synchronizing coefficient (or "elem. In general, the QV loop is much faster than the Pf loop, due to the mechanical inertia constants in the latter. If it can be assumed that the transients in the QV loop are essentially over before the Pf loop reacts, then the coupling between loops can be neglected.

This is a reasonable assumption, and we shall adopt it in the following analysis, due to the resulting simplicity.

THERMAL POWER GENERATION

According to the Annual Report 1995-96 of the Ministry of Coal, Government of India, the coal reserves of Orissa is 46,722 million tonnes out of 2,01,953 million tones estimated for the whole country. Around 85 per cent of the coal reserves in the country are in the non-cooking coal category. The non-cooking coal has ash content of 30-50 per cent with considerable amount of extraneous content.

The deposit of coal in Orissa is mostly concentrated in Talcher and Ib Valley area, which is around 23 per cent of the entire coal potential in the country. This can sustain the power generation of the State as well as a source of power supply to the rest of the country.

The aspects related to thermal generation are as follows.

- Availability of power grade coal and feasibility of its transportation
- Availability of sources of water
- Availability of land for power plant, township and ash disposal
- The geological suitability of sites, soil conditions and amenability to floods
- Environmental considerations
- Evacuation of power to load centre

The generation potential for the coal reserve could be gauged from the fact that a 300 MW thermal plant running at PLF of 75 per cent consumes around 13 million tones of coal per annum. Therefore, the coal deposit in Orissa can sustain 100,000 MW of generation for a period of 100 years.

State Thermal	Installed Capacity	Firm Power
TTPs	460	255
OPGC	420	260
Central Thermal	Orissa Share	
Farakka (1600MW)	235	143
Kahalgaon (840MW)	135	82
Kaniha Stage-I (2*500 MW)	262	160
Total Thermal	1512	900

In addition to this major CPPs like National Aluminium Company, Rourkela Steel Plant, Indian Charge Chrome Ltd, Chowdar, Ferroo Alloys Corporation, Bhadrak and other Captive Power Plants have total installed capacity of 1380

MW. The CEA with the approval of Central Government have published a document viz. fuel of India where the broad scenario of power generation by 2012 has been brought out which indicates the following thermal power stations for consideration in the State of Orissa

Name of Proposed Power Stations	Installed Capacity	Year of Completion (proposed)
Hirma Power Project developed	6x660MW=3960 MW (stage-I)	2002-07
CEPA	6x660MW=3960 MW (stage-I)	2007-12
AES Ib Valley Unit 5 & 6	2x250 MW=500 MW	2002-07
OPGC Unit – 3 & 4	2x210 MW=420 MW	2002-07
Duburi KPCL0	2x250 MW=500 MW	2002-07
Naraj	2x250 MW=500 MW	2002-07
NTPC, Kalimela	2x500 MW=1000 MW	2002-07
	2x500 MW=1000 MW	2007-12
	Total	11,840 MW

The addition of these mega power station will not only meet the requirement of the State but shall be a source of the power starved State of the other region.

HYDRO POTENTIAL

Among the states of the Eastern Region Orissa is favourably placed as far as the hydro resources are concerned .

State Hydro-electricity Project	Installed Capacity	Firm Power
Hirakud	331.5	134
Balimela	360	135
Rengali	250	60
Upper Kolab	320	95
Indravati	600	224
Machkund (Orissa share)	34.43	34
Total Hydro	1895.93	682

The future planning involves installation of new hydel projects to meet the future demands of the State and this is given below .

Balimela	150 MW
Hirakud-B (4x52 MW)	208 MW
Chipilima-13 (4x50 MW)	200 MW
Sindol-I (5x20 MW)	100 MW
Sindol-II (5x20 MW)	100 MW
Sindol-III (2x60 MW)	120 MW
Balimela Dam Toe (2x30 MW) 50% Orissa share	60 MW
Total	938 MW

In addition to this there is also scope of development a number of small, mini and micro hydel power stations to the extent of 200 MW among with the following are worth mentioning.

- Samal Barrage – 180MW

- Jalput Dam Toe – 18 MW (50 % Orissa share)
- Harabhangi – 24 MW
- Salandi – 6 MW
- Bargarh canal – 6 MW
- Lower Machkund – 25 MW
- Lower Kolab – 20 MW

PERSPECTIVE TRANSMISSION PLAN

A report on the Perspective Transmission Plan 2011-12 was prepared by the Central Electricity Authority in June 1999 and released by the Hon'ble Minister of Power on the 1st October 1999. A number of transmission layouts have been proposed corresponding to the two generation scenarios considered in the Report, so that, at an appropriate time in the future, with the uncertainty in generation capacity addition reduced, and with the views of important power sector agencies available, a nationally agreed direction for the development of the National Grid would be formulated, from amongst the indicated and other options. Certain issues which pertain to generation and transmission system development in the 2011-12 time frame have been highlighted in the Perspective Transmission Plan 2011-12, and in relation to these issues, a number of options are still open. Regional level consultations and national level consultations are proposed to be held so that a nationally agreed direction for the development of the National Grid could be formulated.

APPROACHES TO THE PERCEPTIVE TRANSMISSION PLAN

The basic approach to the Perspective Transmission Plan 2011-12 has been that the country as a whole would be treated as one spatial unit for planning with free flow of power across regional boundaries or state boundaries. The generational schemes would be developed keeping in view the suitability of generation sites and the preferences expressed by the promoters, in particular the Independent Power Producers.

The identification of the bulk transmission corridors has been done in such way that even if there are changes with regard to the exact location of certain large thermal power stations, a transmission corridor would not be changed altogether. With this end in view, power in the exporting regions has been sought to be pooled at a few locations and then transmitted by extra high voltage AC or High Voltage Direct Current (HVDC) transmission lines of high capacity to the importing region and then depooled in the importing region for transmission to smaller load centres.

The bulk power transmission system associated with mega power projects has been sought to be integrated with the bulk power transmission system for inter-regional power transfers. In this way the right of way requirements have been sought to be conserved.

POWER EXCHANGE SCENARIOS

Two power exchange scenarios have been considered in the Perspective Transmission Plan 2011-12 upto 10500 MW flow from one region to another has been considered. These exchanges are based on the generating capacity projected for 2011-12. Mostly Northern region will have deficit of power which will be largely met from thermal generating stations planned in Orissa.

There are two generation scenarios for 2011-12, and the ratio of hydroelectric capacity to thermal (including nuclear) capacity in one is different from the corresponding ratio in the other. In Generation Scenario-I, 69,359 MW hydroelectric capacity and 1,71,563 MW thermal (including nuclear) capacity have been considered. In Generation Scenario-II, the capacities considered are 77,948 MW and 1,62,935 MW respectively.

One of the important parameters on the basis of which the national power highways for bulk inter-regional transfer would be identified is the expected quantum of inter-regional power exchange. Therefore, in the light of reviews of generating capacity expansion plans and power surveys, the projections for installed generating capacity in the various regions, the load projections in the various regions, and the corresponding quota of power exchange between the regions are required to be reviewed in the light of new developments.

GENERATING CAPACITY IN THE SOUTHERN REGION AND THE TALCHER-KOLAR HIGH VOLTAGE DIRECT CURRENT BIPOLE

The surplus power in the Southern Region in the various despatches under the two generation scenarios has been taken to be between 1000 MW, and 6000 MW, as indicated in the preceding section. This projection is based on the expectations that a number of generating stations would be constructed in the Southern Region based on imported coal /LNG both under station/Central Sector and by Independent Power Producers. At present there is a huge shortage of electrical energy in the Southern Region and the Talcher Stage-II (4 × 500 MW) project in the Eastern Region has been dedicated to the Southern Region with power transfer planned on the Talcher - Kolar High Voltage Direct Current (HVDC) bipole of 2000 MW capacity.

TRANSIENT STABILITY IN POWER SYSTEM

The ability of a synchronous power system to return to stable condition and maintain its synchronism following a relatively large disturbance arising from very general situations like switching 'on' and 'off' of circuit elements, or clearing of faults etc. is referred to as the transient stability in power systemof the system. More often than not, the power generation systems are subjected to faults of this kind, and hence its extremely important for power engineers to be well-versed with the stability conditions of the system. In general practice studies related to transient stability in power system are done over a very small period of time equal to the time required for one swing, which approximates to around 1 sec or even less. If the system is found to be stable during this first swing, its assumed that the disturbance will reduce in the subsequent swings, and the system will be stable thereafter as is generally the case. Now in order to mathematically determine whether a system is stable or not we need to derive the swing equation of power system.

SWING EQUATION FOR DETERMINING TRANSIENT STABILITY

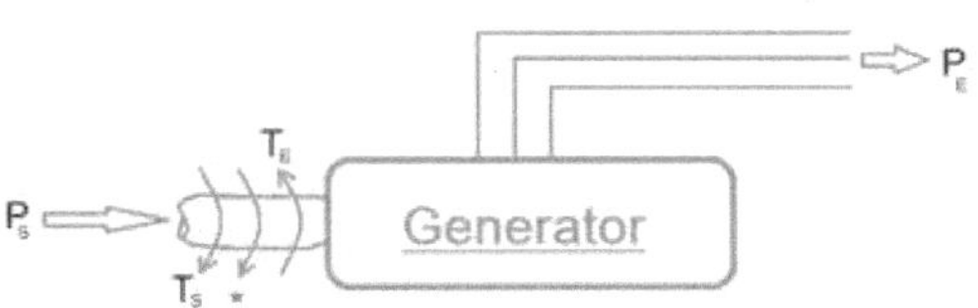

In order to determine the transient stability of a power system using swing equation, let us consider asynchronous generatorsupplied with input shaft power P_s producing mechanical torque equal to T_s.

This makes the machine rotate at a speed of ù rad/sec and the output electromagnetic torque and power generated on the receiving end are expressed as T_e and P_e respectively.

When the synchronous generator is fed with a supply from one end and a constant load is applied to the other, there is some relative angular displacement between the rotor axis and the stator magnetic field, known as the load angle ä which is directly proportional to the loading of the machine. The machine at this instance is considered to be running under stable condition.

Now if we suddenly add or remove load from the machine the rotor decelerates or accelerates accordingly with respect to the stator magnetic field. The operating condition of the machine now becomes unstable and the rotor is now said to be swinging w.r.t the stator field and the equation we so obtain giving the relative motion of the load angle ä w.r.t the stator magnetic field is known as the swing equation for transient stability of power system. Here for the sake of understanding we consider the case where a synchronous generator is suddenly applied with an increased amount of electromagnetic load, which leads to instability by making P_E less than P_S as the rotor undergoes deceleration.

Now the increased amount of the accelerating power required to bring the machine back to stable condition is given by,

$$\text{Accelerating power}, P_{AG} = P_S - P_E$$

Similarly, the accelerating torque is given by,

$$T_{AG} = T_S - T_E$$

Now we know that

$$P_{AG} = T_{AG}\omega = I\alpha\omega$$

(since T = current X angular acceleration)
Furthermore, angular momentum, M = I ù

$$P_{AG} = M\alpha$$

But since on loading the angular displacement è varies continuously with time, we can write.

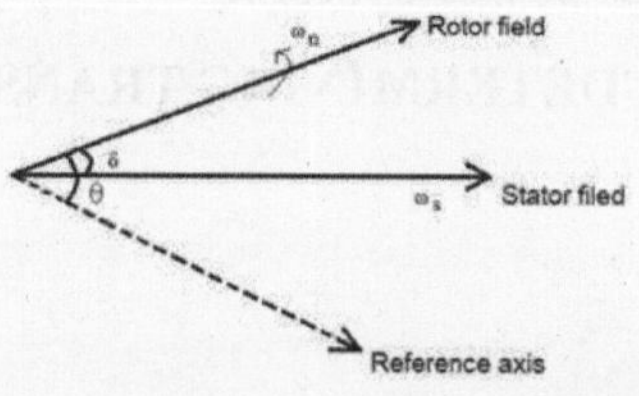

Angular position of rotor with respect to reference axis.

$\bullet = \bullet_s + d\bullet/dt$
Double differentiating w.r.t time, we get,
$d^2\bullet/dt^2 = d^2\bullet/dt^2$
where angular acciletation $\bullet = d^2\bullet/dt^2 = d^2\bullet/dt^2$
Thus we can write,
$P_{AG} = M\, d^2\bullet/dt^2$
Or $M\, d^2\bullet/dt^2 = P_S - P_E$
Now the electromagnetic power transmitted is given by,
$P_E = {}^{V}{}_{G} V_M/{}_{X} \sin \bullet = P_{max} \sin \bullet$
Since when $\bullet = 0$, maximum amplitude $= {}^{V}{}_{G} V_M/{}_{X}$
Thus we can write,
$M\, d^2\bullet/dt2 = P_S - P_{max}$ Sin ä
This is known as the swing equation for transient stability in power system.

POWER-ANGLE RELATIONSHIP

The power-angle relationship. We shall consider this relation for a lumped parameter lossless transmission line. Consider the single-machine-infinite-bus(SMIB) system. In this the reactance *X* includes the reactance of the transmission line and the synchronous reactance or the transient reactance of the generator. The sending end voltage is then the internal emf of the generator. Let the sending and receiving end voltages be given by

$$V_S = V_1\angle\delta,\ V_R = V_2\angle 0^\circ$$

Fig.: An SMIB system.

We then have

$$I_S = \frac{V_1 \angle\delta - V_2}{jX} = \frac{V_1 \cos\delta - V_2 + jV_1 \sin\delta}{jX}$$

The sending end real power and reactive power are then given by

$$P_s + jQ_S = V_S I_S^+ = V_1 (\cos\delta + j\sin\delta)\frac{V_1 \cos\delta - V_2 - jV_1 \sin\delta}{-jX}$$

This is simplified to

$$P_S + jQ_S = \frac{V_1 V_2 \sin\delta + j(V_1^2 - V_1 V_2 \cos\delta)}{X}$$

Since the line is loss less, the real power dispatched from the sending end is equal to the real power received at the receiving end. We can therefore write

$$P_e = P_S = P_R = \frac{V_1 V_2}{X}\sin\delta = P_{max} \sin\delta$$

where $P_{max} = V_1 V_2 / X$ is the maximum power that can be transmitted over the transmission line. The power-angle curve. For a given power P_0. There are two possible values of the angle ä - ä$_0$ and ä$_{max}$. The angles are given by

$$\delta_0 = \sin^{-1}\left(\frac{P_0}{P_{max}}\right)$$

$$\delta_{max} = 180° - \delta_0$$

SWING EQUATION

Let us consider a three-phase synchronous alternator that is driven by a prime mover. The equation of motion of the machine rotor is given by

$$J\frac{d^2\theta}{dt^2} = T_m - T_e = T_a$$

where

J is the total moment of inertia of the rotor mass in kgm^2

T_m is the mechanical torque supplied by the prime mover in N-m

T_e is the electrical torque output of the alternator in N-m

è is the angular position of the rotor in rad

Neglecting the losses, the difference between the mechanical and electrical torque gives the net accelerating torque T_a. In the steady state, the electrical torque is equal to the mechanical torque, and hence the accelerating power will be zero. During this period the rotor will move at synchronous speed ù$_s$ in rad/s. The angular position è is measured with a stationary reference frame. To represent it with respect to the synchronously rotating frame, we define

$$\theta = \omega_s t + \delta$$

where *ä* is the angular position in rad with respect to the synchronously rotating reference frame. Taking the time derivative of the above equation we get

$$I_S = \frac{V_1\angle\delta - V_2}{jX} = \frac{V_1\cos\delta - V_2 + jV_1\sin\delta}{jX}$$

Defining the angular speed of the rotor as

$$\omega_r = \frac{d\theta}{dt}$$

we can write (above) as

$$\omega_r - \omega_s = \frac{d\delta}{dt}$$

We can therefore conclude that the rotor angular speed is equal to the synchronous speed only when *dä/ dt* is equal to zero. We can therefore term *dä/ dt* as the error in speed. Taking derivative of (above equation), we can then rewrite (above equation) as

$$J = \frac{d^2\delta}{dt^2} = T_m - T_e = T_a$$

Multiplying both side of (above equation) by $\grave{u}_m$ we get

$$J\omega_r = \frac{d^2\delta}{dt^2} = T_m - T_e = T_a$$

where $P_{m'}$ P_e and P_a respectively are the mechanical, electrical and accelerating power in MW.

We now define a normalized inertia constant as

$$H\frac{\text{Stored kinetic energy at synchronou s speed in mega} - \text{joules}}{\text{Generator MVA rating}} = \frac{J\omega_s^2}{2S_{rated}}$$

Substituting (above equations) we get

$$2H\frac{S_{rated}}{\omega_s^2}\omega_r\frac{d^2\delta}{dt^2} = P_m - P_e = P_a$$

In steady state, the machine angular speed is equal to the synchronous speed and hence we can replace $\grave{u}_r$ in the above equation by $\grave{u}_s$. Note that in $P_{m'}$ P_e and P_a are given in MW. Therefore dividing them by the generator MVA rating S_{rated} we can get these quantities in per unit. Hence dividing both sides of by S_{rated} we get $\frac{2H}{\omega_s}\frac{d^2\delta}{dt^2} = P_m - P_e = P_a$ per unit

The behaviour of the rotor dynamics and hence is known as the swing equation. The angle *ä* is the angle of the internal emf of the generator and it dictates the amount of power that can be transferred. This angle is therefore called the load angle .

EQUAL AREA CRITERION

The real power transmitted over a lossless line. Now consider the situation in which the synchronous machine is operating in steady state delivering a power P_e equal to P_m when there is a fault occurs in the system. Opening up of the circuit breakers subsequently clears the fault. The circuit breakers take about 5/6 cycles to open and the subsequent post-fault transient last for another few cycles. The input power, on the other hand, is supplied by a prime mover that is usually driven by a steam turbine. The time constant of the turbine mass system is of the order of few seconds, while the electrical system time constant is in milliseconds. Therefore, for all practical purpose, the mechanical power is remains constant during this period when the electrical transients occur. The transient stability study therefore concentrates on the ability of the power system to recover from the fault and deliver the constant power P_m with a possible new load angle ä .

Consider the power angle curve. Suppose the system is operating in the steady state delivering a power of P_m at an angle of $ä_0$ when due to malfunction of the line, circuit breakers open reducing the real power transferred to zero. Since P_m remains constant, the accelerating power P_a becomes equal to P_m. The difference in the power gives rise to the rate of change of stored kinetic energy in the rotor masses. Thus the rotor will accelerate under the constant influence of non-zero accelerating power and hence the load angle will increase. Now suppose the circuit breaker re-closes at an angle $ä_c$. The power will then revert back to the normal operating curve. At that point, the electrical power will be more than the mechanical power and the accelerating power will be negative. This will cause the machine decelerate. However, due to the inertia of the rotor masses, the load angle will still keep on increasing. The increase in this angle may eventually stop and the rotor may start decelerating, otherwise the system will lose synchronism.

Note that

$$\frac{d}{dt}\left(\frac{d\delta}{dt}\right)^2 = 2\left(\frac{d\delta}{dt}\right)\left(\frac{d^2\delta}{dt^2}\right)$$

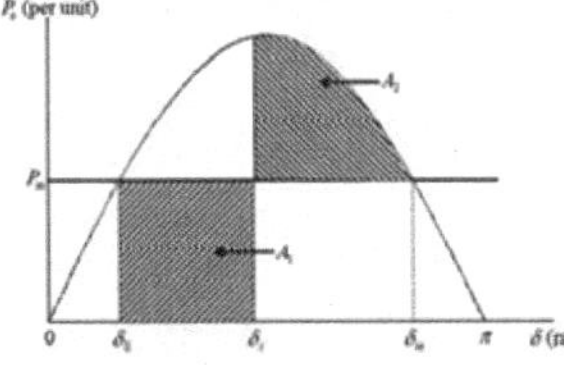

Fig. Power-angle curve for equal area criterion.

Hence multiplying both sides by $d\delta / dt$ and rearranging we get

$$\frac{H}{\omega_s}\frac{d}{dt}\left(\frac{d\delta}{dt}\right)^2 = (P_m - P_e)\frac{d\delta}{dt}$$

Multiplying both sides of the above equation by *dt* and then integrating between two arbitrary angles $\ddot{a}_0$ and $\ddot{a}_c$ we get

$$\frac{H}{\omega_s}\left(\frac{d\delta}{dt}\right)^2\Bigg|_{5_0}^{5_r} = \int_{5_0}^{5_r}(P_m - P_e)d\delta$$

Now suppose the generator is at rest at $\ddot{a}_0$. We then have $d\ddot{a}/dt = 0$. Once a fault occurs, the machine starts accelerating. Once the fault is cleared, the machine keeps on accelerating before it reaches its peak at $\ddot{a}_c$, at which point we again have $d\ddot{a}/dt = 0$. Thus the area of accelerating is given from (above equation) as

$$A_1 = \int_{\delta_0}^{\delta_r}(P_m - P_e)d\delta = 0$$

In a similar way, we can define the area of deceleration. The area of acceleration is given by A_1 while the area of deceleration is given by A_2. This is given by

$$A_2 = \int_{\delta_0}^{\delta_m}(P_e - P_m)d\delta = 0$$

Now consider the case when the line is reclosed at $\ddot{a}_c$ such that the area of acceleration is larger than the area of deceleration, *i.e.*, $A_1 > A_2$. The generator load angle will then cross the point $\ddot{a}_m$, beyond which the electrical power will be less than the mechanical power forcing the accelerating power to be positive. The generator will therefore start accelerating before is slows down completely and will eventually become unstable.

If, on the other hand, $A_1 < A_2$, *i.e.*, the decelerating area is larger than the accelerating area, the machine will decelerate completely before accelerating again.

The rotor inertia will force the subsequent acceleration and deceleration areas to be smaller than the first ones and the machine will eventually attain the steady state.

If the two areas are equal, *i.e.*, $A_1 = A_2$, then the accelerating area is equal to decelerating area and this is defines the boundary of the stability limit. The clearing angle $\ddot{a}_c$ for this mode is called the Critical Clearing Angle and is denoted by $\ddot{a}_{cr}$. By substituting $\ddot{a}_c = \ddot{a}_{cr}$

$$\int_{\delta_0}^{\delta_m}(P_m - P_e)d\delta = \int_{\delta_m}^{\delta_m}(P_e - P_m)d\delta$$

We can calculate the critical clearing angle from the ab move equation. Since the critical clearing angle depends on the equality of the areas, this is called the equal area criterion.

Since we are interested in finding out the maximum time that the circuit breakers may take for opening, we should be more concerned about the critical clearing time rather than clearing angle.

Furthermore, notice that the clearing angle is independent of the generalised inertia constant H . Hence we can comment that the critical clearing angle in this case is true for any generator that has a d-axis transient reactance of 0.20 per unit. The critical clearing time, however, is dependent on H and will vary as this parameter varies.

To obtain a description for the critical clearing time, let us consider the period during which the fault occurs. We then have $P_e = 0$. We can therefore write from

$$\frac{d^2\delta}{dt^2} = \frac{\omega_s}{2H}P_m$$

Integrating the above equation with the initial acceleration being zero we get

$$\frac{d\delta}{dt} = \int_0^t \frac{\omega_s}{2H}P_m dt = \frac{\omega_s}{2H}P_m t$$

Further integration will lead to

$$\delta = \int_0^t \frac{\omega_s}{2H}P_m t dt = \frac{\omega_s}{4H}P_m t^2 + \delta_0$$

Replacing $\ddot{a}$ by $\ddot{a}_{cr}$ and t by t_{cr} in the above equation, we get the critical clearing time as

$$t_{a} = \sqrt{\frac{4H}{\omega_5 P_m}(\delta_a - \delta_0)}$$

To illustrate the response of the load angle $\ddot{a}$, the swing equation is simulated in MATLAB. The swing equation is then expressed as

$$\frac{d\Delta\omega_r}{dt} = \frac{1}{2H}\left(P_m - P_e\right)$$

$$\frac{d\delta}{dt} = \omega_5 \times \Delta\omega_r$$

where Äù$_r$ is the deviation for the rotor speed from the synchronous speed ù$_s$. It is to be noted that the swing equation does not contain any damping. Usually a damping term, that is proportional to the machine speed Äù$_r$, is added with the accelerating power.

Without the damping the load angle will exhibit a sustained oscillation even when the system remains stable when the fault cleared within the critical clearing time.

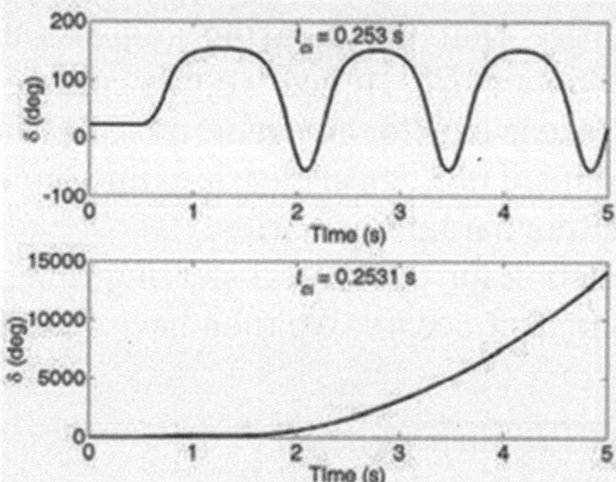

Fig. Stable and unstable system response as a function of clearing time.

The response of the load angle ä for two different values of load angle. It is assumed that the fault occurs at 0.5 s when the system is operating in the steady state delivering 0.9 per unit power. The load angle during this time is constant at 23.96° . The load angle remains stable, albeit the sustained oscillation when the clearing time t_{cl} is 0.253 s. The clearing angle during this time is 88.72° . The system however becomes unstable when the clearing time 0.2531s and the load angle increases asymptotically. The clearing time in this case is 88.77° . This is called the Loss of Synchronism. It is to be noted that such increase in the load angle is not permissible and the protection device will isolate the generator from the system. The clearing time is derived based on the assumption that the electrical power P_e becomes zero during the fault. This need not be the case always. In that even we have to resort to finding the clearing time using the numerical integration of the swing equation.

OBJECTIVE OF POWER SYSTEM PROTECTION

The objective of power system protection is to isolate a faulty section of electrical power system from rest of the live system so that the rest portion can function satisfactorily without any severer damage due to fault current.

Actually circuit breaker isolates the faulty system from rest of the healthy system and this circuit breakers automatically open during fault condition due to its trip signal comes from protection relay. The main philosophy about protection is that no protection of power system can prevent the flow of faultcurrent through the system, it only can prevent the continuation of flowing of fault current by quickly disconnect the short circuit path from the system. For satisfying this quick disconnection the protection relays should have following functional requirements.

PROTECTION SYSTEM IN POWER SYSTEM

Let's have a discussion on basic concept of protection system in power system and coordination of protection relays.

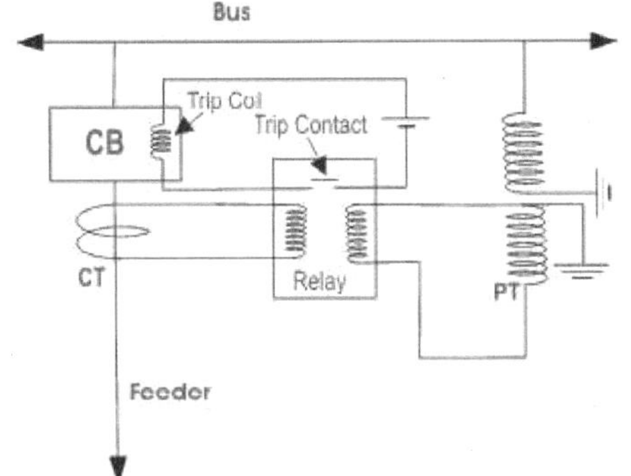

Basic connection diagram of protection relay

In the picture the basic connection of protection relay has been shown. It is quite simple. The secondary of currenttransformer is connected to the current coil of relay.

And secondary ofvoltage transformer is connected to the voltagecoil of the relay. Whenever any fault occurs in the feeder circuit, proportionate secondary current of the CT will flow through the current coil of the relay due to which mmf of that coil is increased.

This increased mmf is sufficient to mechanically close the normally open contact of the relay.

This relay contact actually closes and completes the DC trip coil circuit and hence the trip coil is energized. The mmf of the trip coil initiates the mechanical movement of the tripping mechanism of the circuit breaker and ultimately the circuit breaker is tripped to isolate the fault.

RELIABILITY

The most important requisite of protective relay is reliability. They remain inoperative for a long time before a fault occurs; but if a fault occurs, the relays must respond instantly and correctly.

SELECTIVITY

The relay must be operated in only those conditions for which relays are commissioned in the electrical power system. There may be some typical condition during fault for which some relays should not be operated or operated after some definite time delay hence protection relay must be sufficiently capable to select appropriate condition for which it would be operated.

SENSITIVITY

The relaying equipment must be sufficiently sensitive so that it can be operated reliably when level of fault condition just crosses the predefined limit.

SPEED

The protective relays must operate at the required speed. There must be a correct coordination provided in various power system protection relays in such a way that for fault at one portion of the system should not disturb other healthy portion.

Fault current may flow through a part of healthy portion since they are electrically connected but relays associated with that healthy portion should not be operated faster than the relays of faulty portion otherwise undesired interruption of healthy system may occur.

Again if relay associated with faulty portion is not operated in proper time due to any defect in it or other reason, then only the next relay associated with the healthy portion of the system must be operated to isolate the fault. Hence it should neither be too slow which may result in damage to the equipment nor should it be too fast which may result in undesired operation.

SWITCHGEAR

Consists of mainly bulk oil circuit breaker, minimum oil circuit breaker, SF_6circuit breaker, air blast circuit breaker and vacuum circuit breaker etc. Different operating mechanisms such as solenoid, spring, pneumatic, hydraulic etc. are employed in circuit breaker. Circuit breaker is the main part of protection system in power system it automatically isolate the faulty portion of the system by opening its contacts.

PROTECTIVE GEAR

Consists of mainly power system protection relays like current relays, voltagerelays, impedance relays, power relays, frequency relays, etc. based on operating parameter, definite time relays, inverse time relays, stepped relays etc. as per operating characteristic, logic wise such as differential relays, over fluxing relays etc. During fault the protection relay gives trip signal to the associated circuit breaker for opening its contacts.

STATION BATTERY

All the circuit breakers of electrical power system are DC (Direct Current) operated. Because DC power can be stored in battery and if situation comes when total failure of incoming power occurs, still the circuit breakers can be operated for restoring the situation by the power of storage battery . Hence thebattery .is another essential item of the power system. Some time it is referred as the heart of the electrical substation. An electrical substation battery or simply a station battery containing a number of cells accumulate energy during the period of availability of A.C supply and discharge at the time when relays operate so that relevant circuit breaker is tripped.

PROTECTION SYSTEM IN POWER SYSTEM

Let's have a discussion on basic concept of protection system in power system and coordination of protection relays.

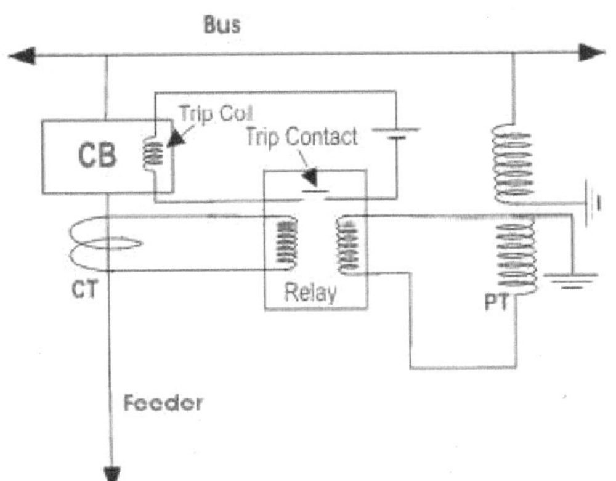

Basic connection diagram of protection relay

In the picture the basic connection of protection relay has been shown. It is quite simple. The secondary of current transformer is connected to the current coil of relay. And secondary of voltage transformer is connected to the voltage coil of the relay.

Whenever any fault occurs in the feeder circuit, proportionate secondary current of the CT will flow through the current coil of the relay due to which mmf of that coil is increased.

This increased mmf is sufficient to mechanically close the normally open contact of the relay. This relay contact actually closes and completes the DC trip coil circuit and hence the trip coil is energized.

The mmf of the trip coil initiates the mechanical movement of the tripping mechanism of the circuit breaker and ultimately the circuit breaker is tripped to isolate the fault.

FUNCTIONAL REQUIREMENTS OF PROTECTION RELAY

Reliability

The most important requisite of protective relay is reliability. They remain inoperative for a long time before a fault occurs; but if a fault occurs, the relays must respond instantly and correctly.

Selectivity

The relay must be operated in only those conditions for which relays are commissioned in the electrical power system. There may be some typical condition during fault for which some relays should not be operated or operated after some definite time delay hence protection relay must be sufficiently capable to select appropriate condition for which it would be operated.

Sensitivity

The relaying equipment must be sufficiently sensitive so that it can be operated reliably when level of fault condition just crosses the predefined limit.

Speed

The protective relays must operate at the required speed. There must be a correct coordination provided in various power system protection relays in such a way that for fault at one portion of the system should not disturb other healthy portion.

Fault current may flow through a part of healthy portion since they are electrically connected but relays associated with that healthy portion should not be operated faster than the relays of faulty portion otherwise undesired interruption of healthy system may occur.

Again if relay associated with faulty portion is not operated in proper time due to any defect in it or other reason, then only the next relay associated with the healthy portion of the system must be operated to isolate the fault.

Hence it should neither be too slow which may result in damage to the equipment nor should it be too fast which may result in undesired operation.

IMPORTANT ELEMENTS FOR POWER SYSTEM PROTECTION

Switchgear

Consists of mainly bulk oil circuit breaker, minimum oil circuit breaker, SF_6 circuit breaker, air blast circuit breaker and vacuum circuit breaker etc. Different operating mechanisms such as solenoid, spring, pneumatic, hydraulic etc. are employed in circuit breaker. Circuit breaker is the main part of protection system in power system it automatically isolate the faulty portion of the system by opening its contacts.

Protective Gear

Consists of mainly power system protection relays like current relays, voltage relays, impedance relays, power relays, frequency relays, etc. based on operating parameter, definite time relays, inverse time relays, stepped relays etc. as per operating characteristic, logic wise such as differential relays, over fluxing relays etc. During fault the protection relay gives trip signal to the associated circuit breaker for opening its contacts.

Station Battery

All the circuit breakers of electrical power system are DC (Direct Current) operated. Because DC power can be stored in battery and if situation comes when total failure of incoming power occurs, still the circuit breakers can be operated for restoring the situation by the power of storage battery. Hence the battery is another essential item of the power system. Some time it is referred as the heart of the electrical substation. An electrical substation battery or simply a station battery containing a number of cells accumulate energy during the period of availability of A.C supply and discharge at the time when relays operate so that relevant circuit breaker is tripped.

5

Renewable Energy

Contents

- Little to No Global Warming Emissions
- Improved Public Health and Environmental Quality
- A Vast and Inexhaustible Energy Supply
- Jobs and Other Economic Benefits
- Stable Energy Prices
- A More Reliable and Resilient Energy System

Renewable energy — wind, solar, geothermal, hydroelectric, and biomass — provides substantial benefits for our climate, our health, and our economy:

Each source of renewable energy has unique benefits and costs; this page explores the many benefits associated with these energy technologies.

Little to No Global Warming Emissions

Human activity is overloading our atmosphere with carbon dioxide and other global warming emissions, which trap heat, steadily drive up the planet's temperature, and create significant and harmful impacts on our health, our environment, and our climate.

Electricity production accounts for more than one-third of U.S. global warming emissions, with the majority generated by coal-fired power plants, which produce approximately 25 percent of total U.S. global warming emissions; natural gas-fired power plants produce 6 percent of total emissions. In contrast, most renewable energy sources produce little to no global warming emissions. According to data aggregated by the International Panel on Climate Change, life-cycle global warming emissions associated with renewable energy—including manufacturing, installation, operation and maintenance, and dismantling and decommissioning—are minimal.

Compared with natural gas, which emits between 0.6 and 2 pounds of carbon dioxide equivalent per kilowatt-hour (CO2E/kWh), and coal, which emits between 1.4 and 3.6 pounds of CO2E/kWh, wind emits only 0.02 to 0.04 pounds of CO2E/kWh, solar 0.07 to 0.2,geothermal 0.1 to 0.2, and hydroelectric between 0.1 and 0.5. Renewable electricity generation from biomass can have a wide range of global warming emissions depending on the resource and how it is harvested. Sustainably sourced biomass has a low emissions footprint, while unsustainable sources of biomass can generate significant global warming emissions.

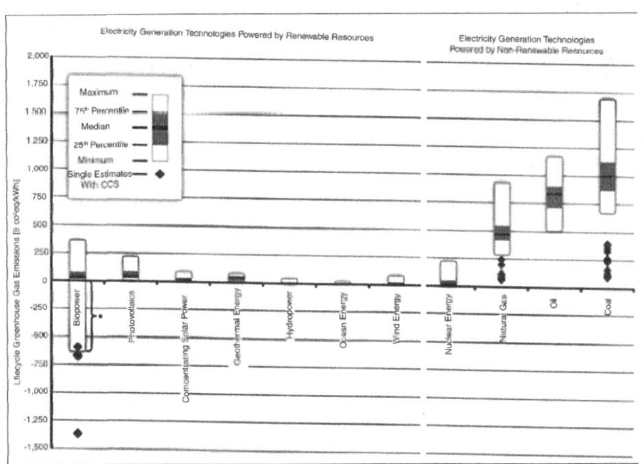

Source: IPCC, 2011: IPCC Special Report on Renewable Energy Sources and Climate Change Mitigation. Prepared by Working Group III of the Intergovernmental Panel on Climate Change.

Increasing the supply of renewable energy would allow us to replace carbon-intensive energy sources and significantly reduce U.S. global warming emissions. For example, a 2009 UCS analysis found that a 25 percent by 2025 national renewable electricity standard would lower power plant CO2 emissions 277 million metric tons annually by 2025—the equivalent of the annual output from 70 typical (600 MW) new coal plants. In addition, a ground-breaking study by the U.S. Department of Energy's National Renewable Energy Laboratory explored the feasibility and environmental impacts associated with generating

80 percent of the country's electricity from renewable sources by 2050 and found that global warming emissions from electricity production could be reduced by approximately 81 percent.

Improved Public Health and Environmental Quality

Generating electricity from renewable energy rather than fossil fuels offers significant public health benefits. The air and water pollution emitted by coal and natural gas plants is linked to breathing problems, neurological damage, heart attacks, and cancer.

Replacing fossil fuels with renewable energy has been found to reduce premature mortality and lost workdays, and it reduces overall health care costs.

The aggregate national economic impact associated with these health impacts of fossil fuels is between $361.7 and $886.5 billion, or between 2.5 percent and 6 percent of gross domestic product (GDP).

Wind, solar, and hydroelectric systems generate electricity with no associated air pollution emissions. While geothermaland biomass energy systems emit some air pollutants, total air emissions are generally much lower than those of coal- and natural gas-fired power plants.

In addition, wind and solar energy require essentially no water to operate and thus do not pollute water resources or strain supply by competing with agriculture, drinking water systems, or other important water needs. In contrast, fossil fuels can have a significant impact on water resources.

For example, both coal mining and natural gas drilling can pollute sources of drinking water. Natural gas extraction by hydraulic fracturing (fracking) requires large amounts of water and all thermal power plants, including those powered by coal, gas, and oil, withdraw and consume water for cooling.

Biomass and geothermal power plants, like coal- and natural gas-fired power plants, require water for cooling. In addition, hydroelectric power plants impact river ecosystems both upstream and downstream from the dam. However, NREL's 80 percent by 2050 renewable energy study, which included biomass and geothermal, found that water withdrawals would decrease 51 percent to 58 percent by 2050 and water consumption would be reduced by 47 percent to 55 percent.

Throughout the United States, strong winds, sunny skies, plant residues, heat from the earth, and fast-moving water can each provide a vast and constantly replenished energy resource supply. These diverse sources of renewable energy have the technical potential to provide all the electricity the nation needs many times over.

Renewable Resource	**Electricity Generation CapacityPotential (gigawatts)**	**Electricity Generation Potential (Billionn Kilowatt-Hours)**	**Renewable Electricity Generation as Percent of 2012 Electricity use**
Wind			
Land-Based	10,955	32,784	809%
Offshore	4,223	16,976	419%
Sybtotal	15,178	49,760	1,227%
Solar			
Photovoltaics	154,856	283,664	6,997%
Concentrating Solar Power	38,066	116,146	2,865%
Subtotal	192,922	399,810	9,862%
Bioenergy			
Subtotal	62	488	12%
Geothermal			
Hydrothermal	38	308	8%
Enhanced Geothermal	3,976	31,345	773%

Systems			
Subtotal	4,014	31,653	781%
Hydropower			
Existing Conventional	78	277	7%
New Conventional	60	259	6%
Subtotal 1	38	536	13%
Total	212,314	482,247	11,896%

Estimates of the technical potential of each renewable energy source are based on their overall availability given certain technological and environmental constraints. In 2012, NREL found that together, renewable energy sources have the technical potential to supply 482,247 billion kilowatt-hours of electricity annually. This amount is 118 times the amount of electricity the nation currently consumes. However, it is important to note that not all of this technical potential can be tapped due to conflicting land use needs, the higher short-term costs of those resources, constraints on ramping up their use such as limits on transmission capacity, barriers to public acceptance, and other hurdles. Today, renewable energy provides only a tiny fraction of its potential electricity output in the United States and worldwide. But numerous studies have repeatedly shown that renewable energy can be rapidly deployed to provide a significant share of future electricity needs, even after accounting for potential constraints.

Jobs and Other Economic Benefits

Compared with fossil fuel technologies, which are typically mechanized and capital intensive, the renewable energy industry is more labour-intensive. This means that, on average, more jobs are created for each unit of electricity generated from renewable sources than from fossil fuels.

Renewable energy already supports thousands of jobs in the United States. For example, in 2011, the wind energy industry directly employed 75,000 full-time-equivalent employees in a variety of capacities, including manufacturing, project development, construction and turbine installation, operations and maintenance, transportation and logistics, and financial, legal, and consulting services.

More than 500 factories in the United States manufacture parts for wind turbines, and the amount of domestically manufactured equipment used in wind turbines has grown dramatically in recent years: from 35 percent in 2006 to 70 percent in 2011.

Other renewable energy technologies employ even more workers. In 2011, the solar industry employed approximately 100,000 people on a part-time or full-time basis, including jobs in solar installation, manufacturing, and sales; the hydroelectric power industry employed approximately 250,000 people in 2009; and in 2010 the geothermal industry employed 5,200 people.

Increasing renewable energy has the potential to create still more jobs. In 2009, the Union of Concerned Scientists conducted an analysis of the economic benefits of a 25 percent renewable energy standard by 2025; it found that such a policy would create more than three times as many jobs as producing an equivalent amount of electricity from fossil fuels—resulting in a benefit of 202,000 new jobs in 2025.

In addition to the jobs directly created in the renewable energy industry, growth in renewable energy industry creates positive economic "ripple" effects. For example, industries in the renewable energy supply chain will benefit, and unrelated local businesses will benefit from increased household and business incomes. In addition to creating new jobs, increasing our use of renewable energy offers other important economic development benefits. Local governments collect property and income taxes and other payments from renewable energy project owners.

These revenues can help support vital public services, especially in rural communities where projects are often located. Owners of the land on which wind projects are built also often receive lease payments ranging from $3,000 to $6,000 per megawatt of installed capacity, as well as payments for power line easements and road rights-of-way. Or they may earn royalties based on the project's annual revenues. Similarly, farmers and rural landowners can generate new sources of supplemental income by producing feedstocks for biomass power facilities.

UCS analysis found that a 25 by 2025 national renewable electricity standard would stimulate $263.4 billion in new capital investment for renewable energy technologies, $13.5 billion in new landowner income biomass production and/ or wind land lease payments, and $11.5 billion in new property tax revenue for local communities.

Renewable energy projects therefore keep money circulating within the local economy, and in most states renewable electricity production would reduce the need to spend money on importing coal and natural gas from other places. Thirty-eight states were net importers of coal in 2008—from other states and, increasingly, other countries: 16 states spent a total of more than $1.8 billion on coal from as far away as Colombia, Venezuela, and Indonesia, and 11 states spent more than $1 billion each on net coal imports.

Stable Energy Prices

Renewable energy is providing affordable electricity across the country right now, and can help stabilize energy prices in the future.

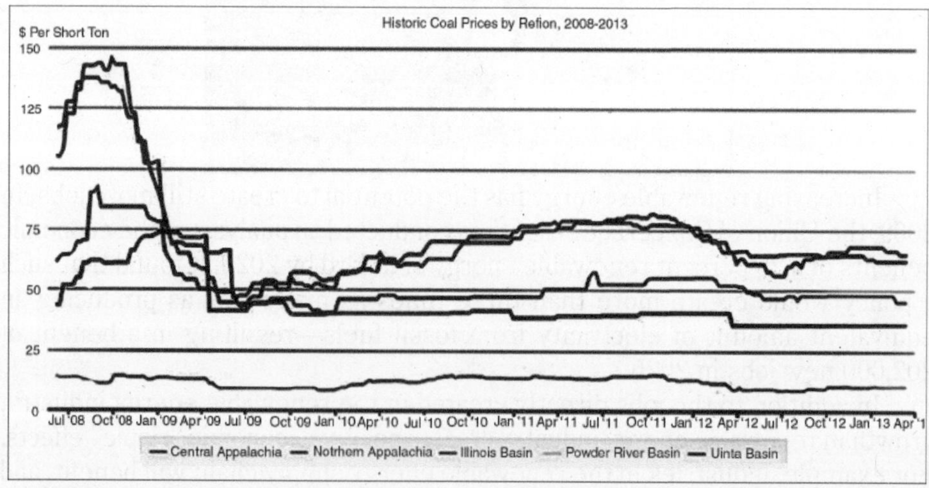

The costs of renewable energy technologies have declined steadily, and are projected to drop even more. For example, the average price of a solar panel has dropped almost 60 percent since, 2011 . The cost of generating electricity from wind dropped more than 20 percent between 2010 and 2012 and more than 80 percent since, 1980. In areas with strong wind resources like

Texas, wind power can compete directly with fossil fuels on costs. The cost of renewable energy will decline even further as markets mature and companies increasingly take advantage of economies of scale. While renewable facilities require upfront investments to build, once built they operate at very low cost and, for most technologies, the fuel is free. As a result, renewable energy prices are relatively stable over time. UCS's analysis of the economic benefits of a 25 percent renewable electricity standard found that such a policy would lead to 4.1 percent lower natural gas prices and 7.6 percent lower electricity prices by 2030.

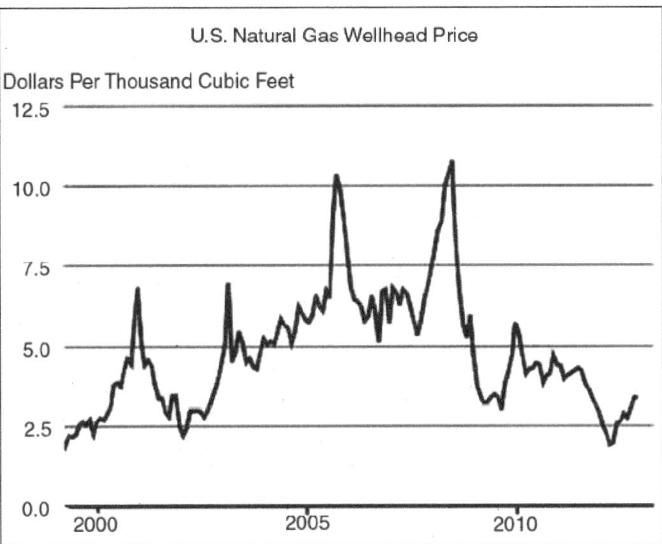

In contrast, fossil fuel prices can vary dramatically and are prone to substantial price swings. For example, there was a rapid increase in U.S. coal prices due to rising global demand before 2008, then a rapid fall after 2008 when global demands declined. Likewise, natural gas prices have fluctuated greatly since, 2000. Using more renewable energy can lower the prices of and demand for natural gas and coal by increasing competition and diversifying our energy supplies. An increased reliance on renewable energy can help protect consumers when fossil fuel prices spike. In addition, utilities spend millions of dollars on financial instruments to hedge themselves from these fossil fuel price uncertainties. Since, hedging costs are not necessary for electricity generated from renewable sources, long-term renewable energy investments can help utilities save money they would otherwise spend to protect their customers from the volatility of fossil fuel prices.

Wind and solar are less prone to large-scale failure because they are distributed and modular. Distributed systems are spread out over a large geographical area, so a severe weather event in one location will not cut off power to an entire region. Modular systems are composed of numerous individual wind turbines or solar arrays. Even if some of the equipment in the system is damaged, the rest can typically continue to operate. For example, in 2012 Hurricane Sandy damaged fossil fuel-dominated electric generation and distribution systems in New York and New Jersey and left millions of people without power. In contrast, renewable energy projects in the Northeast weathered Hurricane Sandy with minimal damage or disruption.

A More Reliable and Resilient Energy System

The risk of disruptive events will also increase in the future as droughts, heat waves, more intense storms, and increasingly severe wildfires become more frequent due to global warming. Renewable energy sources are more resilient than coal, natural gas, and nuclear power plants in the face of these sorts of extreme weather events.

For example, coal, natural gas, and nuclear power depend on large amounts of water for cooling, and limited water availability during a severe drought or heat wave puts electricity generation at risk. Wind and solar photovoltaic systems do not require water to generate electricity, and they can help mitigate risks associated with water scarcity.

RENEWABLE ENERGY DEVELOPMENT IN INDIA

India has done a significant progress in the power generation in the country. The installed generation capacity was 1300 megawatt (MW) at the time of Independence *i.e.,* about 60 year's back. The total generating capacity anticipated at the end of the Tenth Plan on 31-03-2007, is 1, 44,520 MW which includes the generation through various sectors like Hydro, Thermal and Nuclear. The power generation in the country is planned through funds provided by the Central Sector, State Sector and Private Sector. The power shortages noticed is of the order of 11 per cent. In the opinion of the experts such short fall can be reduced through proper management and thus almost 40 per cent energy can be saved. It has been noticed that one watt saved at the point of consumption is more than 1.5 watts generated. In terms of Investment it costs around ₹.40 million to generate one MW of new generation plant, but if the same ₹.40 million is spent on conservation of energy methods, it can provide up to 3 MW of avoidable generation capacity.

There are about 80,000 villages yet to be electrified for which provision has been made to electrify 62,000 villages from grid supply in the Tenth Plan. It is planned that participation of decentralized power producers shall be ensured, particularly for electrification of remote villages in which village level organisations shall play a crucial role for the rural electrification programme.

Emphasis is given to the renewable energy programme towards gradual commercialisation. This programme is looked after by the Ministry of Non-Conventional Sources of energy. Simultaneously private sector investments in renewable energy sources are also increased to promote power generation. So far an excessive reliance was preferred on the use of fossil fuel resources like coal, oil and natural gas to meet the power requirement of the country which was not suitable in the long run due to limited availability of the fossil fuel as well as the adverse impact on the environment and ecology.

Since, the availability of fossil fuel is on the decline therefore, in this backdrop the norms for conventional or renewable sources of energy (RSE) is given importance not only in India but has attracted the global attention.

The main items under RSE are as follows:

- Hydro Power
- Solar Power
- Wind Power
- Bio-mass Power
- Energy from waste
- Ocean energy
- Alternative fuel for surface transportation

HYDRO POWER

India is endowed with a large potential of hydro power, of which only 17 per cent has been harnessed so far. The hydro electricity is a clean and renewable source of energy. It has been felt that there is a long gestation period in hydro projects due to delays in forest and environment clearance, rehabilitation of the project effected people besides inter-state disputes and construction holdups due to several reasons. Under RSE only small hydro projects are considered since, they do not require large pondage and have the capacity to provide power to remote and hilly terrain where extension of the grid system is either un-economic or not possible. It has been estimated that the potential available in the country under small hydropower schemes is of the order of 15000 MW in which the plans that are considered are up to 25 MW capacity individually which are classified as small hydro projects under the Ministry of Non-Conventional Sources of energy. The small hydro power stations are mostly located in hilly areas and are given priority for local benefits to the residents which provide them gainful employment through the energy potential.

SOLAR POWER:-

The climatic condition in India provides abundant potential of solar power due to large scale radiation available during a wider part of the year due to tropical condition in the country. The solar power can be developed for long term use through the application of solar photo- voltaic (SPV) Technology which provides a potential of 20MW per sq. Km. The other method for Utilisation of solar energy is through the adoption of solar thermal Technology. The programmes are under way to utilize SPV by connecting to grid power systems.

It has come to notice from a report of Xinhva news agency that Shanghai, the business capital of China, is launching a 100000 rooftop solar photo voltaic (SPV) system which would generate 430 million KWH of electricity which would be enough to supply power to the entire city for two days.

The other popular use is by stand- alone applications which include solar powered street lights, domestic lights, water pumps etc. The cost of the photo SPV modules is quite expensive which is in the range of $ 3-4 per watt, in spite of best efforts, the price could not come down in India, China and other countries. The effort is to bring the price down to $ 1 per watt when it may be more popular for use. The efforts to use amorphous silicon technology were cheaper but its long terms use is not practicable. The SPV technology if cheep, would be useful for people living in far - flung areas as extending grid would involve high cost.

The solar thermal devices are widely used in the country for various purposes such as solar water heaters, solar cookers, solar dryers etc. There is wide scope for development of solar thermal application for which the research is in progress. The energy obtained through Solar Thermal route is 35 MW per sq. km.

WIND POWER

The wind power development in the country is largely of recent period which has been found to be quite impressive. As per available data, it is 5340 MW by March 31, 2006, through wind power. Earlier it was estimated that the potential for wind power in the country was 20,000 MW which has been revised to 45000MW after collecting the data on the potential available in the coastal and other areas of the country. At present India is fifth in the world after Germany, USA, Denmark and Spain in terms of wind power. It has been observed that the private sector is showing interest in setting of wind power projects. The unit size of wind turbine generators which were earlier in the range of 55-100 kw are now preferred in the range of 750-1000 kw. It has been observed that the productivity of the larger machine is higher as compared to the smaller machine. In respect of cost consideration, it has been noticed that the cost of such a project is about ₹.40 million to ₹.50 million per MW which includes all local civil, electrical works and erection also. The life of a wind power project is estimated to be about 20 years.

China has guaranteed all certified renewable energy producers in its service area that the grid will purchase their power and the price will be spread out to all the users across the grid. According to sources, such commitments can only spur further development in the renewable energy sector.

BIO-MASS POWER

There is quite a high energy potential available in the country in resources such as firewood, agroresidues and animal wastes. These resources are mainly utilized by the rural population of the country. It has been estimated that there is a potential to install 19500 MW capacity through biomass conservation technologies like combustion, gasification, incineration and also bagasse – based co- generation in sugar mills. So far only around 380 MW of this potential has been tapped and there is wide scope for expanding the size of their use for the benefit of the majority of the rural population to meet their energy needs.

ENERGY FROM WASTE

It has been estimated that there is about 30 million tones by solid waste and 4400 million cubic meters of liquid waste generated every year in urban areas through domestic as well as commercial establishment. The manufacturing sector also contributes high quantity of waste. It has been estimated that through garbage there is a potential to generate 1700 MW of electricity. However all these activities are still to be given a practical shape.

OCEAN ENERGY

The Ocean on the earth covers about 71 per cent of the total surface which collects and store solar energy. If this energy is quantified in terms of Oil, it can be said that an amount of solar radiation equivalent in heat content to about 245 billion barrels of oil is absorbed by the sea. The energy available in the Ocean is clean, continuous and renewable. In future it would be possible to tap energy from the sea.

ALTERNATIVE FUEL FOR SURFACE TRANSPORTATION:

Hydrocarbons used as fuels for transportation are to be replaced by other eco-friendly fuels for surface transport vehicles. Many options such as compressed natural gas (CNG), battery – powered vehicles and fuel cells are currently available. The use of diesel in transportation in Delhi was causing pollution in the air. The Government has adopted CNG use for all vehicles using diesel fuel, which has improved the environment significantly

POLICY FOLLOWED IN CHINA.

Renewable energy is very much promoted by the Chinese Government. Earlier the emphasis was quite low but of late it has been observed that a high priority is given to renewable sources of energy. The new law stipulates the responsibilities of government and society in developing and applying renewable energy.

At the same time as the law was passed, the Chinese Government set a target for renewable energy to contribute 10 per cent of the country's gross energy consumption by 2020, a huge increase from the current 1 per cent Seeing this as a future stimulus of renewable energy development, many Chinese and international observers are very excited, expecting a tremendous growth in the renewable energy market in the next 15 years as the result of the implementation of this law.

The policy adopted also provides subsidies and tax credits. However, wind power development in China has lagged far behind the world leaders in the past 10 years The total installed capacity only hit 769 MWrecently, whereas, in India, the second largest developing country, the total capacity is 5340 MW. Between 1999 and 2002, only 211 MW was installed in China, while globally, 18,000 MW was added during the same period.

TRANSFORMER TECHNOLOGY

Transformers were first used in India in 1897 to light Darjeeling Municipal area. The commercial production of transformers commenced. in 1936 at Government Electrical factory, Bangalore. Later on new companies have started production and improvement in the Transformer Technology. When transformer factories were set up in 1960s there were no vendors in the country to supply processed raw materials and accessories.

Evolution of power transformer technology in the country during the past five decades is quite impressive.

There are manufacturers in the country with full access to the latest technology at the global level. Some of the manufacturers have impressive R&D set up to support the technology.

AMORPHOUS METAL DISTRIBUTION TRANSFORMER

Huge amount of energy has been lost due to no-load loss of transformer core. Energy efficient transformer such as AMDT can effectively lower such energy wastage and at the same time reduce green house gases emission.

Reliability in distribution system can be brought about by incorporating following steps.

- Use transformers, which have minimum maintenance problems.
- Improve power factor of the system.
- Ensure proper protection to the system.
- Neutral grounding system should be effective.
- Introduce maintenance free equipment like Vacuum Circuit Breakers for all 11 KV feeders with auto re-closers.
- Undertake preventive maintenance and avoid emergencies.

GLOBAL WARMING AND CLIMATE CHANGE

It has been felt that there is rising demand for energy, food and raw materials by a population of 2.5 billion Chinese and Indians. Both these countries have large coal dominated energy systems in the world and the use of fossil fuels such as coal and oil releases carbon dioxide (Co_2) into the air which adds to the greenhouse gases which lead to global warming. At present US is the largest contributor of Co_2 emissions but the development in India and China is going to increase their share in emission of such a gas. According to Kyoto Protocol this has to be controlled. Climate change shall be a cause of extinction of many bird varieties and other animals on the earth.

Renewable source of energy is the best solution for such a problem in the world. Both India and China are trying to develop their technology in this regard. India has the world's fourth largest wind power industry, while China is the global leader in harnessing solar energy for hot water.

Wind Power could generate almost 29 percent of the world's electricity by 2030 and was growing faster than any other clean energy source, a wind business group and environmental lobby Greenpeace said. 'At good locations wind can compete with the cost of both coal and gas-fired Power' the Global Wind Energy Council (GWEC) and Greenpeace said in a study, 'Global Wind Energy Outlook 2006'.The two said that wind, which now accounts for 0.8 percent of the world's electricity supply, was expanding faster than other renewable energies such as solar, geothermal or tidal power in a shift from fossil fuels.

There have been cases of farmers committing suicides due to poverty and failure of crop in some parts of India. A World Bank study released has found a correlation between climate change and farmer suicides. It says poor farmers who are unable to adapt to changing climates fall into debt and later, death traps.

It can be surmised that energy development should be preferable by adopting measures which does not give rise to greenhouse gasses as it would effect change in climate leading to overall difficulties to the people who are accustomed to the climate as prevailing on the earth.

6

Synchronous Motors

PRINCIPLE OF OPERATION

INTRODUCTION

Synchronous motors have the same construction as alternators. The few special features relative to the production of the direct current necessary for their excitation will be treated separately, later. It will be assumed that the reader is already familiar with the general details of construction of alternators.

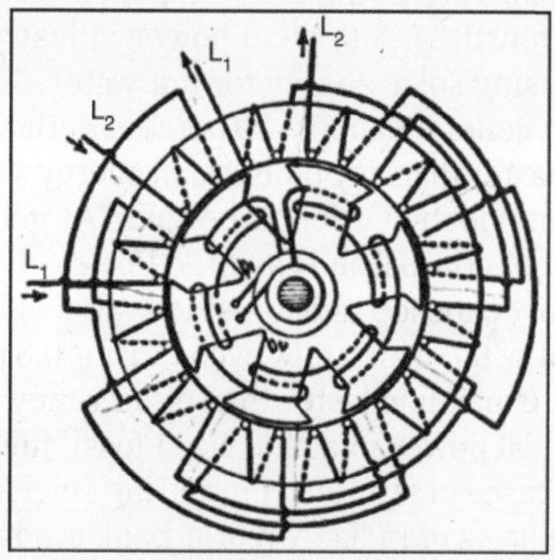

There are motors having movable armatures and stationary fields, or vice versa, and also motors with revolving iron masses in which all the windings are stationary.

These machines are similar to the generators of the same types; for example, Fig. indicates, diagrammatically, the principle of construction of a two-phase synchronous motor, with a ring armature and movable fields, receiving an exciting current through the brushes b_1 and b_2.

These motors are designed like generators, the essential condition to be fulfilled being to have a low armature-reaction and powerful inducing fields, in order to obtain good stability.

NUMBER OF POLES

Although it is more difficult to increase the number of poles for small powers than for large powers, the construction of small synchronous motors for ordinary frequencies (40 to 60 cycles) presents no special difficulties, if the speeds corresponding to these frequencies are not objectionable, because these speeds are perfectly allowable so far as centrifugal force is concerned.

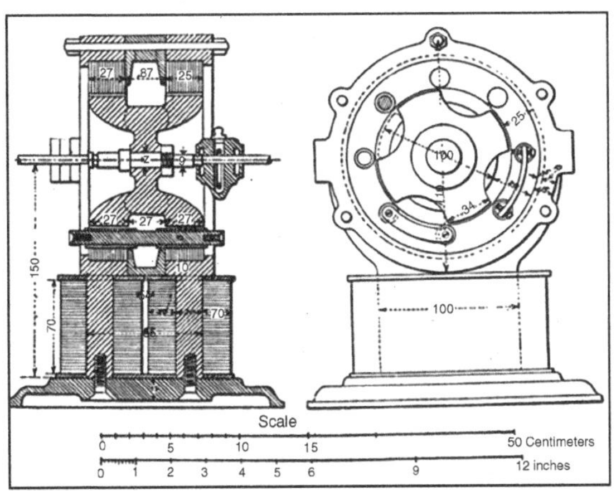

SYNCHRONOUS MOTORS AT LOW SPEEDS

On the other hand, in the construction of small synchronous motors to run at low angular velocities, it is extremely difficult to find space for the numerous conductors and for the exciting or field coils, which must produce as many ampere-turns as in the case of large motors. For this reason non-synchronous motors are more convenient for low rotative speeds.

The author has been able, however, to produce motors of low power (a few hundred watts) which have moving iron and have a very high number of poles (as many as 50 for example), by utilizing inductiontype excitation, the magnetic circuit being closed exteriorly, as shown in Fig, in such a way as to allow all the space needed for the exciting coils. These coils can then be replaced by permanent magnets, thus producing motors which run without excitation, at speeds sufficiently low to be synchronized by hand, and which can render useful service, in certain applications, such as for oscillographs.

For this purpose the author preferably employs a small horseshoe magnet that is made to revolve around a stationary armature having a number of poles which is a multiple of 6. It is possible, in this way, to obtain very stable synchronous rotation of a revolving mirror without expending more than 1.5 to 2 watts.

Several firms made a specialty of synchronous motors, at an early date, among which we may mention La Societe 1'Eclairage Electrique in France, and the Fort Wayne Company in America.

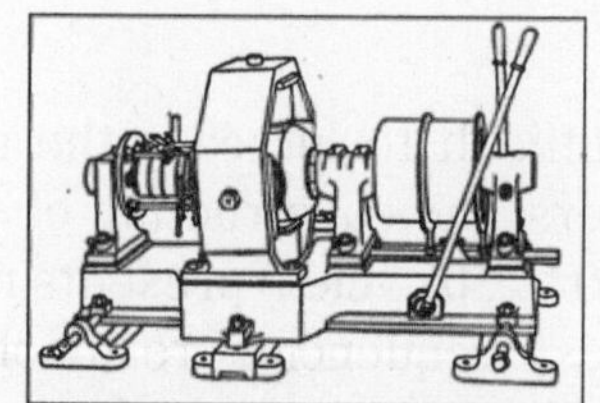

One form of motor constructed in France by the Societe 1'Eclairage Electrique, is constructed for polyphase currents or for single-phase currents, for powers ranging from 1 to 130 H.P.

APPLICATION OF SYNCHRONOUS MOTORS

Synchronous machines are very important machinery in electrical engineering.

Following are the important applications of a synchronous motors or machines:

- It is used in power houses and sub-stations in parallel to the bus bars to improve power factor. For this purpose it is run without mechanical load on it and over excited.
- In factories having large number of induction motors or transformers operating at lagging power factor, it is used for improving power factor.
- It is used to generate electric power at power station, one of the most important application of synchronous machines.
- it is used to control the voltage at the end of transmission line by varying its excitation.
- It is also used in rubber mills, textile mills, cement factories, air compressors, centrifugal pumps which requiring constant speed.
- It is used in motor genreator sets requiring constant speed.

Synchronous motors are mostly use to drive continuously operating and constant speed equipment such as centrifugal pumps, fans, blowers, ammonia and air compressors, motor-generator sets etc.

ADVANTAGES

Being power factor improvement appliance synchronous machines posses lots of positive points.

Some very important advantages are:

- There motors can be made to operate at a leading power factor and thereby improve the power factor of an industrial plant from laggoi to one that is close to unity.
- It gives constant speed from no load to full load.
- Electro-magnetic power varies linearly with voltage.
- These motors can be constructed with wider air gaps than induction motors, which make then better mechanically.

- These motors operates at higher efficiency, especially in the low speed unity power factor range.

DISADVANTAGES

Everything has its own advantages and disadvantages and Synchronous machine is not an exception. Synchronous machines has some disadvantages.

Some of noticeable disadvantages or demarits of synchronous machines are as:

- It cannot be used for variable speed job as there is no possibility of speed adjustment.
- It requires d. c. excitation which must be supplied from external source.
- It cannot be started under loaded condition. Its starting torque is zero.
- It has a tendency to hunt.
- It is not possible for places where frequent starting is required.
- It may fall out of synchronism and stop when over loaded.
- Collector rings and brushes are required.

COMPARISON BETWEEN SYNCHRONOUS MOTOR AND INDUCTION MOTORS

1. Synchronous motors operate at synchronous speed (RPM=120f/p) while induction motors operate at less than synchronous speed (RPM=120f/p– slip). Slip is nearly zero at zero load torque and increases as load torque increases.
2. Synchronous motors require DC excitation to be supplied to the rotor windings; induction motors don't.
3. Synchronous motors require a DC power source for the rotor excitation.
4. Synchronous motors require slip rings and brushes to supply rotor excitation. Induction motors don't require slip rings, but some induction motors have them for soft starting or speed control.
5. Synchronous motors require rotor windings while induction motors are most often constructed with conduction bars in the rotor that are shorted together at the ends to form a "squirrel cage."
6. Synchronous motors require a starting mechanism in addition to the mode of operation that is in effect once they reach synchronous speed. Three phase induction motors can start by simply applying power, but single phase motors require an additional starting circuit.
7. The power factor of a synchronous motor can be adjusted to be lagging, unity or leading while induction motors must always operate with a lagging power factor.

8. Synchronous motors are generally more efficient that induction motors.
9. Synchronous motors can be constructed with permanent magnets in the rotor eliminating the slip rings, rotor windings, DC excitation system and power factor adjustability.
10. Synchronous motors are usually built only is sizes larger than about 1000 Hp (750 kW) because of their cost and complexity. However, permanent magnet synchronous motors and electronically controlled permanent synchronous motors called brushless DC motors are available in smaller sizes.

STARTING OF DC MOTOR

In any electric motor, operation is based on simple electromagnetism. A current-carrying conductor generates a magnetic field; when this is then placed in an external magnetic field, it will experience a force proportional to the current in the conductor, and to the strength of the external magnetic field.

As you are well aware of from playing with magnets as a kid, opposite (North and South) polarities attract, while like polarities (North and North, South and South) repel. The internal configuration of a DC motor is designed to harness the magnetic interaction between a current-carrying conductor and an external magnetic field to generate rotational motion.

WORKING

Let's start by looking at a simple 2-pole DC electric motor (here red represents a magnet or winding with a "North" polarization, while green represents a magnet or winding with a "South" polarization).

Every DC motor has six basic parts— axle, rotor (a.k.a., armature), stator, commutator, field magnet(s), and brushes. In most common DC motors (and all that Beamers will see), the external magnetic field is produced by high-strength permanent magnets.

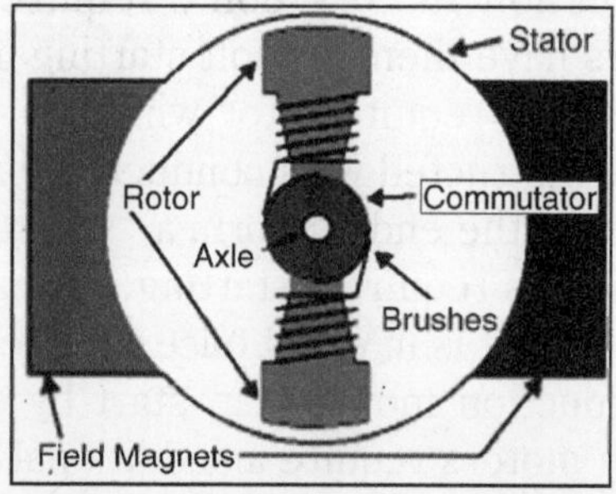

The stator is the stationary part of the motor— this includes the motor casing, as well as two or more permanent magnet pole pieces. The rotor (together with the axle and attached commutator) rotate with respect to the stator. The rotor consists of windings (generally on a core), the windings being electrically connected to the commutator. The above diagram shows a common motor layout–with the rotor inside the stator (field) magnets.

The geometry of the brushes, commutator contacts, and rotor windings are such that when power is applied, the polarities of the energized winding and the stator magnet(s) are misaligned, and the rotor will rotate until it is almost aligned with the stator's field magnets. As the rotor reaches alignment, the brushes move to the next commutator contacts, and energize the next winding. Given our example two-pole motor, the rotation reverses the direction of current through the rotor winding, leading to a "flip" of the rotor's magnetic field, driving it to continue rotating.

In real life, though, DC motors will always have more than two poles (three is a very common number). In particular, this avoids "dead spots" in the commutator. You can imagine how with our example two-pole motor, if the rotor is exactly at the middle of its rotation (perfectly aligned with the field magnets), it will get "stuck" there. Meanwhile, with a two-pole motor, there is a moment where the commutator shorts out the power supply (*i.e.,* both brushes touch both commutator contacts simultaneously).

This would be bad for the power supply, waste energy, and damage motor components as well. Yet another disadvantage of such a simple motor is that it would exhibit a high amount of torque "ripple" (the amount of torque it could produce is cyclic with the position of the rotor).

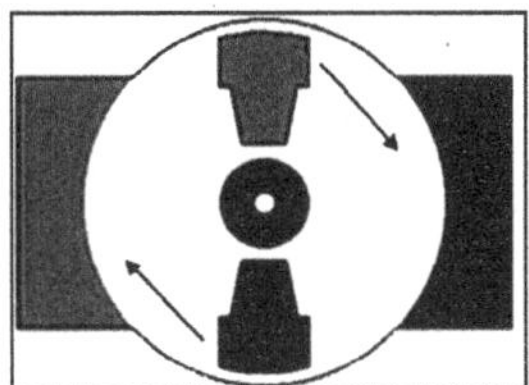

So since most small DC motors are of a three-pole design, let's tinker with the workings of one via an interactive animation (JavaScript required):

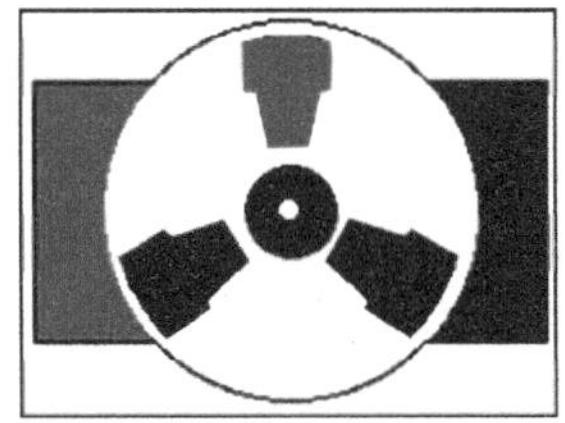

You'll notice a few things from this—namely, one pole is fully energized at a time (but two others are "partially" energized). As each brush transitions from one commutator contact to the next, one coil's field will rapidly collapse, as the next coil's field will rapidly charge up (this occurs within a few microsecond). We'll see more about the effects of this later, but in the meantime you can see that this is a direct result of the coil windings' series wiring:

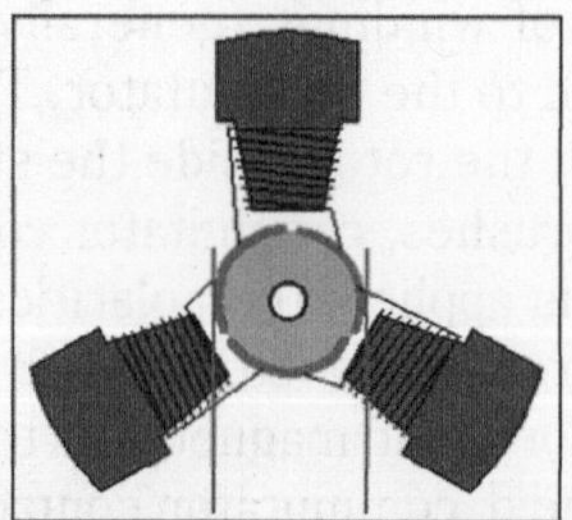

There's probably no better way to see how an average DC motor is put together, than by just opening one up. Unfortunately this is tedious work, as well as requiring the destruction of a perfectly good motor.

Luckily for you, I've gone ahead and done this in your stead. The guts of a disassembled Mabuchi FF-030-PN motor (the same model that Solarbotics sells) are available for you to see here (on 10 lines/ cm graph paper). This is a basic 3-pole DC motor, with 2 brushes and three commutator contacts.

The use of an iron core armature (as in the Mabuchi, above) is quite common, and has a number of advantages. First off, the iron core provides a strong, rigid support for the windings—a particularly important consideration for high-torque motors. The core also conducts heat away from the rotor windings, allowing the motor to be driven harder than might otherwise be the case. Iron core construction is also relatively inexpensive compared with other construction types.

But iron core construction also has several disadvantages. The iron armature has a relatively high inertia which limits motor acceleration. This construction also results in high winding inductances which limit brush and commutator life.

In small motors, an alternative design is often used which features a 'coreless' armature winding.

This design depends upon the coil wire itself for structural integrity. As a result, the armature is hollow, and the permanent magnet can be mounted inside the rotor coil. Coreless DC motors have much lower armature inductance than iron-core motors of comparable size, extending brush and commutator life.

Fig. Diagram Courtesy of MicroMo

The coreless design also allows manufacturers to build smaller motors; meanwhile, due to the lack of iron in their rotors, coreless motors are somewhat prone to overheating. As a result, this design is generally used just in small, low-power motors. BEAMers will most often see coreless DC motors in the form of pager motors.

CHARACTERISTICS OF DC MOTOR

DC motors are divided into three classes, designated according to the method of connecting the armature and the field windings as shunt-series and compound wound.

SHUNT-WOUND MOTORS

This type of motor runs practically constant speed, regardless of the load. It is the type generally used in commercial practice and is usually recommended where starting conditions are not usually severs. Speed of the shunt-wound motors may be regulated in two ways: first, by inserting resistance in series with the armature, thus decreasing speed: and second, by inserting resistance in the field circuit, the speed will vary with each change in load: in the latter, the speeds is practically constant for any setting of the controller. This latter is the most generally used for adjustable-speed service, as in the case of machine tools.

SERIES-WOUND DC MOTORS

This type of motor speed varies automatically with the load, increasing as the load decreases.

Use of series motor is generally limited to case where a heavy power demand is necessary to bring the machine up to speed, as in the case of certain elevator and hoist installations, for steelcars, etc. Series-wound motors should never be used where the motor cab be started without load, since they will race to a dangerous degree.

COMPOUND-WOUND DC MOTORS

A combination of the shunt wound and series wound types combines the characteristics of both. Characteristics may be varied by varying the combination of the two windings. These motors are generally used where severe starting conditions are met and constant speed is required at the same time.

SQUIRREL-CAGE INDUCTION MOTORS

The most simple and reliable of all electric motors. Essentially a constant speed machine, which is adaptable for users under all but the most severe starting conditions. Requires little attention as there are no commutator or slip rings, yet operates with good efficiency.

WOUND-ROTOR (SLIP RING) INDUCTION MOTOR

Used for constant speed-service requiring a heavier starting torque than is obtainable with squirrel cage type. Because of its lower starting current, this type is frequently used instead of the squirrel-cage type in larger sizes.

These motors are also used for varying-speed-service. Speed varies with this load, so that they should not be used where constant speed at each adjustment is required, as for machine tools.

SINGLE PHASE INDUCTION MOTORS

This motor is used mostly in small sizes, where polyphase current is not available. Characteristics are not as good as the polyphase motor and for size larger that 10 HP, the line disturbance is likely to be objectionable. These motors are commonly used for light starting and for running loads up to 1/3 HP Capacitor and repulsion types provide greater torque and are built in sizes up to 10 HP.

SYNCHRONOUS MOTORS

Run at constant speed fixed by frequency of the system. Require direct current for excitation and have low starting torque. For large motor-generators sets, frequency changes, air compressors and similar apparatus which permits starting under a light load, for which they are generally used. These motors are used with considerable advantage, particularly on large power systems, because of their inherent ability to improve the power factor of the system.

USES OF DC MOTOR

DC Motors are argueably the most useful type of electrical motors, and with good reason, they are designed to be used with batteries, solar cells or similar cell based energy sources, and as a result are used in systems where you don't have to be tied to a wall.

Furthermore, even in systems where they are tied to the wall, sometimes it can be more efficient and cost effective to run DC Motors even in certain situations.

AS SHUNT MOTORS

There are three kind of charactersticks for a motor *viz.* Speed-Torque, Speed-Current and Torque-Current charactersticks. After analysing all three charactersticks for D. C. Shunt Motor it is observed that it is an approximately constant speed motor.

It is therefore, used where,

- The speed is required to remain almost constant from no-load condition to full load-condition.

- The load has to be driven at a number of prefer and any one of which is required to remain nearly constant.

Industrial Use: Lathes, Drills, Boring Mills, Shapers, Spinning and Weaving Machines etc.

AS SERIES MOTORS

After analysing all three characterstics for D. C. Series Motor it is observed that it is a variable speed motor. It means speed it low at high torque and *vice-versa*. However, at light or no-load, the motor tends to attain dangerously high speed. The motor has a high starting torque.

It is therefore, used where:

- Large starting torque is required like in Elevators and Electric Traction.
- The load is subjected to heavy fluctuations and the speed is automatically required on sewing machines etc.

Industrial Use: Electric traction, brands, elevators, air compressors, vacuum cleaners, hair drier, sewing machines etc.

AS COMPOUND MOTORS

D. C. Compound Motor are of two types . It is therefore, used where, specification required for particular motor

1. Differential-compound motors are rarely used because of their poor torque characteristics.
2. Cumulative-compound motors are used where a fairly constant speed is required with irregular loads or suddenly applied heavy load .

LOSSES IN DC MACHINE

The losses in a d.c. machine (generator or motor) may be divided into three classes viz

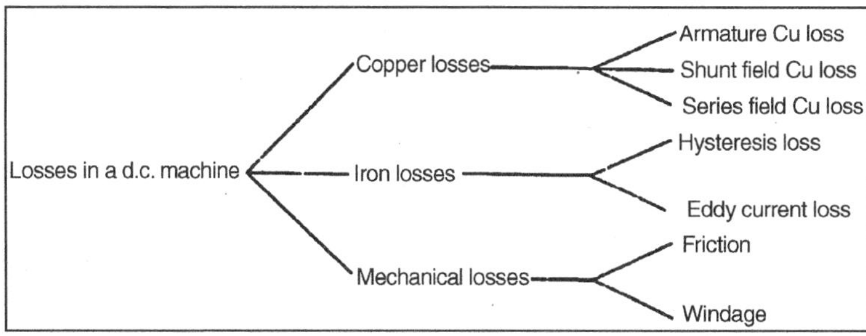

1. Copper losses
2. Iron or core losses and
3. Mechanical losses.

All these losses appear as heat and thus raise the temperature of the machine. They also lower the efficiency of the machine.

COPPER LOSSES

These losses occur due to currents in the various windings of the machine:

1. Armature copper loss = $I_a^2 R_a$
2. Shunt field copper loss = $I_{sh}^2 R_{sh}$
3. Series field copper loss = $I_{se}^2 R_{se}$

There is also brush contact loss due to brush contact resistance (*i.e.,* resistance between the surface of brush and surface of commutator). This loss is generally included in armature copper loss.

IRON OR CORE LOSSES

These losses occur in the armature of a d.c. machine and are due to the rotation of armature in the magnetic field of the poles.

They are of two types viz.,:

1. Hysteresis loss
2. Eddy current loss.

Hysteresis Loss

Hysteresis loss occurs in the armature of the d.c. machine since any given part of the armature is subjected to magnetic field reversals as it passes under successive poles.

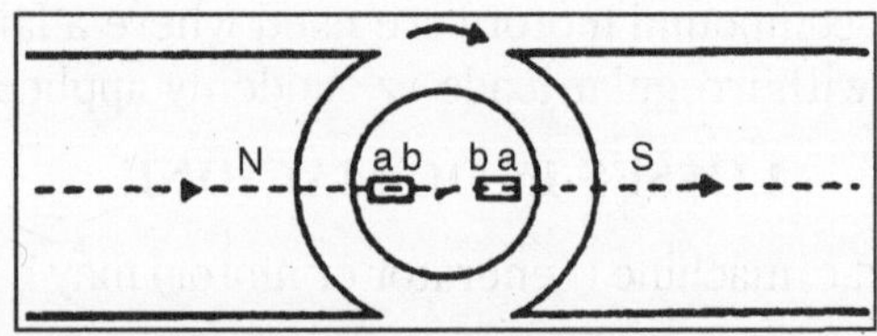

Fig. Shows an Armature Rotating in Two-pole Machine.

Consider a small piece ab of the armature. When the piece ab is under N-pole, the magnetic lines pass from a to b. Half a revolution later, the same piece of iron is under S-pole and magnetic lines pass from b to a so that magnetism in the iron is reversed. In order to reverse continuously the molecular magnets in the armature core, some amount of power has to be spent which is called hysteresis loss. It is given by Steinmetz formula. This formula is

Hysteresis loss,

$$P_h = h\,B_{max}^{16}\,fv \text{ watts}$$

where,

B_{max} = Maximum flux density in armature

f = Frequency of magnetic reversals

= NP/120 where N is in r.p.m.

V = Volume of armature in m^3

η = Steinmetz hysteresis co-efficient

In order to reduce this loss in a d.c. machine, armature core is made of such materials which have a low value of Steinmetz hysteresis co-efficient *e.g.,* silicon steel.

Eddy Current Loss

In addition to the voltages induced in the armature conductors, there are also voltages induced in the armature core. These voltages produce circulating currents in the armature core as shown in Figure. These are called eddy currents and power loss due to their flow is called eddy current loss. The eddy current loss appears as heat which raises the temperature of the machine and lowers its efficiency.

Core resistance can be greatly increased by constructing the core of thin, roundIf a continuous solid iron core is used, the resistance to eddy current path will be small due to large cross-sectional area of the core. Consequently, the magnitude of eddy current and hence eddy current loss will be large. The magnitude of eddy current can be reduced by making core resistance as high as practical. The iron sheets called laminations.

The laminations are insulated from each other with a coating of varnish. The insulating coating has a high resistance, so very little current flows from one lamination to the other. Also, because each lamination is very thin, the resistance to current flowing through the width of a lamination is also quite large. Thus laminating a core increases the core resistance which decreases the eddy current and hence the eddy current loss.

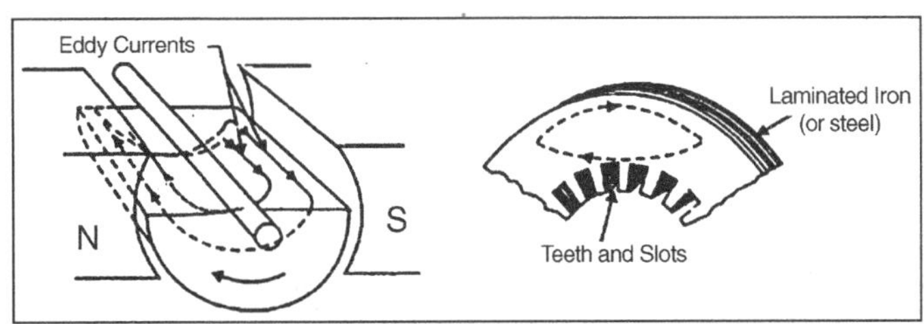

Eddy current loss, $P_e = K_e B_{max}^2 f^2 t^2 V$ watts

where,

Ke = Constant depending upon the electrical resistance of core and system of units used

B_{max} = Maximum flux density in Wb/m^2

f = Frequency of magnetic reversals in Hz

t = Thickness of lamination in m

V = Volume of core in m^3

It may be noted that eddy current loss depends upon the square of lamination thickness. For this reason, lamination thickness should be kept as small as possible.

MECHANICAL LOSSES

These losses are due to friction and windage:

- Friction loss *e.g.,* bearing friction, brush friction etc.
- Windage loss *i.e.,* air friction of rotating armature.

These losses depend upon the speed of the machine. But for a given speed, they are practically constant.

Iron losses and mechanical losses together are called stray losses.

CONSTANT AND VARIABLE LOSSES

The losses in a D.C. generator (or d.c. motor) may be sub-divided into

Constant Losses

Those losses in a d.c. generator which remain constant at all loads are known as constant losses.

The constant losses in a d.c. generator are:

- Iron losses
- Mechanical losses
- Shunt field losses
- Variable losses

Variable Losses

Those losses in a d.c. generator which vary with load are called variable losses.

The variable losses in a d.c. generator are:

- Copper loss in armature winding $\left(I_a^2 R_a\right)$
- Copper loss in series field winding $\left(I_{se}^2 R_{se}\right)$

Total losses = Constant losses + Variable losses

Field Cu loss is constant for shunt and compound generators.

7

Air Standard Cycles

THEORETICAL ANALYSIS

The accurate analysis of the various processes taking place in an internal combustion engine is a very complex problem. If these processes were to be analysed experimentally, the analysis would be very realistic no doubt. It would also be quite accurate if the tests are carried out correctly and systematically, but it would be time consuming. If a detailed analysis has to be carried out involving changes in operating parametres, the cost of such an analysis would be quite high, even prohibitive. An obvious solution would be to look for a quicker and less expensive way of studying the engine performance characteristics. A theoretical analysis is the obvious answer.

A theoretical analysis, as the name suggests, involves analyzing the engine performance without actually building and physically testing an engine. It involves *simulating* an engine operation with the help of thermodynamics so as to formulate mathematical expressions which can then be solved in order to obtain the relevant information. The method of solution will depend upon the complexity of the formulation of the mathematical expressions which in turn will depend upon the assumptions that have been introduced in order to analyse the processes in the engine. The more the assumptions, the simpler will be the mathematical expressions and the easier the calculations, but the lesser will be the accuracy of the final results. The simplest theoretical analysis involves the use of the air standard cycle, which has the largest number of simplifying assumptions.

A THERMODYNAMIC CYCLE

In some practical applications, notably steam power and refrigeration, a thermodynamic cycle can be identified.

A thermodynamic cycle occurs when the working fluid of a system experiences a number of processes that eventually return the fluid to its initial state. In steam power plants, water is pumped (for which work W_P is required) into a boiler and evaporated into steam while heat Q_A is supplied at a high temperature. The steam flows through a turbine doing work W_T and then passes into a condenser where it is condensed into water with consequent rejection of heat Q_R to the atmosphere. Since the water is returned to its initial state, the net change in energy is zero, assuming no loss of water through leakage or evaporation.

AIR STANDARD CYCLES

The air standard cycle is a cycle followed by a heat engine which uses air as the working medium. Since the air standard analysis is the simplest and most idealistic, such cycles are also called *ideal cycles* and the engine running on such cycles are called *ideal engines.*

In order that the analysis is made as simple as possible, certain assumptions have to be made. These assumptions result in an analysis that is far from correct for most actual combustion engine processes, but the analysis is of considerable value for indicating the upper limit of performance. The analysis is also a simple means for indicating the relative effects of principal variables of the cycle and the relative size of the apparatus.

Assumptions

- The working medium is a perfect gas with constant specific heats and molecular weight corresponding to values at room temperature.
- No chemical reactions occur during the cycle. The heat addition and heat rejection processes are merely heat transfer processes.
- The processes are reversible.
- Losses by heat transfer from the apparatus to the atmosphere are assumed to be zero in this analysis.
- The working medium at the end of the process (cycle) is unchanged and is at the same condition as at the beginning of the process (cycle).

In The selecting an idealized process one is always faced with the fact that the simpler the assumptions, the easier the analysis, but the farther the result from reality. The air cycle has the advantage of being based on a few simple assumptions and of lending itself to rapid and easy mathematical handling without recourse to thermodynamic charts or tables or complicated calculations. On the other hand, there is always the danger of losing sight of its limitations and of trying to employ it beyond its real usefulness.

Equivalent Air Cycle

A particular air cycle is usually taken to represent an approximation of some real set of processes which the user has in mind. Generally speaking, the air cycle representing a given real cycle is called an *equivalent air cycle.*

The equivalent cycle has, in general, the following characteristics in common with the real cycle which it approximates:

- A similar sequence of processes.
- Same ratio of maximum to minimum volume for reciprocating engines or maximum to minimum pressure for gas turbine engines.
- The same pressure and temperature at a given reference point.
- An appropriate value of heat addition per unit mass of air.

THE CARNOT CYCLE

This cycle was proposed by Sadi Carnot in 1824 and has the highest possible efficiency for any cycle. Figures show the P-V and T-s diagrams of the cycle.

Assuming that the charge is introduced into the engine at point 1, it undergoes isentropic compression from 1 to 2. The temperature of the charge rises from T_{min} to T_{max}. At point 2, heat is added isothermally. This causes the air to expand, forcing the piston forward, thus doing work on the piston. At point 3, the source of heat is removed and the air now expands isentropically to point 4, reducing the temperature to T_{min} in the process. At point 4, a cold body is applied to the end of the cylinder and the piston reverses, thus compressing the air isothermally; heat is rejected to the cold body. At point 1, the cold body is removed and the charge is compressed isentropically till it reaches a temperature T_{max} once again. Thus, the heat addition and rejection processes are isothermal while the compression and expansion processes are isentropic.

From thermodynamics, per unit mass of charge:

Heat supplied from point 1 to 2

Heat rejected from point 4 to 1

Now $p_2v_2 = RT_{max}$ (7)

And $p_4v_4 = RT_{min}$ (8)

Since Work done, per unit mass of charge, W = heat supplied – heat rejected

We have assumed that the compression and expansion ratios are equal, that is

Heat supplied $Q_s = R\ T_{max} \ln (r)$ (10)

Hence, the thermal efficiency of the cycle is given by

From Equation it is seen that the thermal efficiency of the Carnot cycle is only a function of the maximum and minimum temperatures of the cycle. The efficiency will increase if the minimum temperature (or the temperature at which the heat is rejected) is as low as possible. According to this equation, the efficiency will be equal to 1 if the minimum temperature is zero, which happens to be the absolute zero temperature in the thermodynamic scale.

This equation also indicates that for optimum (Carnot) efficiency, the cycle (and hence the heat engine) must operate between the limits of the highest and lowest possible temperatures. In other words, the engine should take in all the heat at as high a temperature as possible and should reject the heat at as low a temperature as possible. For the first condition to be achieved, combustion (as applicable for a real engine using fuel to provide heat) should begin at the highest possible temperature, for then the irreversibility of the chemical reaction would be reduced. Moreover, in the cycle, the expansion should proceed to the lowest possible temperature in order to obtain the maximum amount of work. These conditions are the aims of all designers of modern heat engines. The conditions of heat rejection are governed, in practice, by the temperature of the atmosphere.

It is impossible to construct an engine which will work on the Carnot cycle. In such an engine, it would be necessary for the piston to move very slowly during the first part of the forward stroke so that it can follow an isothermal process. During the remainder of the forward stroke, the piston would need to move very quickly as it has to follow an isentropic process. This variation in the speed of the piston cannot be achieved in practice. Also, a very long piston stroke would produce only a small amount of work most of which would be absorbed by the friction of the moving parts of the engine.

Since the efficiency of the cycle, as given by Eq. is dependent only on the maximum and minimum temperatures, it does not depend on the working medium. It is thus independent of the properties of the working medium.

Piston Engine Air Standard Cycles

The cycles described here are air standard cycles applicable to piston engines. Engines bases on these cycles have been built and many of the engines are still in use.

The Lenoir Cycle

The Lenoir cycle is of interest because combustion (or heat addition) occurs without compression of the charge. Figures show the P-V and T-s diagrams.

According to the cycle, the piston is at the top dead centre, point 1, when the charge is ignited (or heat is added). The process is at constant volume so the pressure rises to point 2. From 2 to 3, expansion takes place and from 3 to 1 heat is rejected at constant pressure.

Heat supplied, $Q_s = c_v(T_2 - T_1)$ (12)

Heat rejected, $Q_r = c_p(T_3 - T_1)$ (13)

Since $W = Q_s - Q_r$ (14)

$W = c_v(T_2 - T_1) - c_p(T_2 - T_1)$ (15)

Thus,

Here, $r_e = V_3/V_1$, the volumetric expansion ratio. Equation 18 indicates that the thermal efficiency of the Lenoir cycle depends primarily on the expansion ratio and the ratio of specific heats.

The intermittent-flow engine which powered the German V-1 buzz-bomb in 1942 during World War II operated on a modified Lenoir cycle. A few engines running on the Lenoir cycle were built in the late 19th century till the early 20th century.

The Otto Cycle

The Otto cycle, which was first proposed by a Frenchman, Beau de Rochas in 1862, was first used on an engine built by a German, Nicholas A. Otto, in 1876. The cycle is also called a *constant volume* or *explosion* cycle. This is the equivalent air cycle for reciprocating piston engines using spark ignition.

At the start of the cycle, the cylinder contains a mass M of air at the pressure and volume indicated at point:

- The piston is at its lowest position. It moves upward and the gas is compressed isentropically to point
- At this point, heat is added at constant volume which raises the pressure to point
- The high pressure charge now expands isentropically, pushing the piston down on its expansion stroke to point
- Where the charge rejects heat at constant volume to the initial state, point 1.

The Atkinson Cycle

This cycle is also referred to as the *complete expansion cycle.* Inspection of the P-V diagrams of the Otto, Diesel and Dual cycles shows that the expansion process to point 4 does not proceed to the lowest possible pressure, namely, atmospheric pressure. This is true of all real engines; when the exhaust valve opens, the high pressure gases undergo a violent blow down process with consequent dissipation of available energy. This is necessary so as to allow the gases to flow out due to pressure difference and hence reduce the piston work in driving out the gases. The air standard cycle shows a loss of net work because of the reduction in area of the P-V diagram.

In the Otto cycle, if the expansion is allowed to completion to point 4' and heat rejection occurs at constant pressure, the cycle is called the Atkinson cycle.

The heat supplied, Q_s per unit mass of charge is given by,

$$c_{v(}T_3 - T_2)$$

whereas the heat rejected, Q_r per unit mass of charge is given by,

$$c_{p(}T_4 - T_1)$$

Since the Atkinson cycle area under the P-V diagram is larger than the corresponding Otto cycle, the efficiency, for the same compression ratio and heat input, will be higher.

An engine can be built to make use of complete expansion, but the stroke length of such an engine will be extremely long and will not be economically feasible to offset the improvement in power and efficiency. Also, there are some operational problems with such a cycle.

DIESEL AND DUAL CYCLES

DIESEL CYCLE

This cycle, proposed by a German engineer, Dr. Rudolph Diesel to describe the processes of his engine, is also called the constant pressure cycle. This is believed to be the equivalent air cycle for the reciprocating slow speed compression ignition engine. The P-V and T-s diagrams are shown in Figsure respectively. The cycle has processes which are the same as that of the Otto cycle except that the heat is added at constant pressure.

The heat supplied, Q_s is given by $c_p(T_3 - T_2)$

whereas the heat rejected, Q_r is given by $c_v(T_4 - T_1)$

and the thermal efficiency is given by

$$\eta_{th} = 1 - \frac{c_v(T_4 - T_1)}{c_p(T_3 - T_2)}$$

$$= 1 - \frac{1}{\gamma}\left\{\frac{T_1\left(\frac{T_4}{T_1} - 1\right)}{T_2\left(\frac{T_3}{T_2} - 1\right)}\right\}$$

From the T-s diagram, the difference in enthalpy between points 2 and 3 is the same as that between 4 and 1, thus

$$\Delta s_{2-3} = \Delta s_{4-1}$$

$$\therefore c_v \text{In}\left(\frac{T_4}{T_1}\right) = c_p \text{In}\left(\frac{T_3}{T_2}\right)$$

$$\therefore \text{In}\left(\frac{T_4}{T_1}\right) = \gamma \text{In}\left(\frac{T_3}{T_2}\right)$$

$$\therefore \frac{T_4}{T_1} = \left(\frac{T_3}{T_2}\right)^{\gamma} \text{ and } \frac{T_1}{T_2} = \left(\frac{V_2}{V_1}\right)^{\gamma-1} = \frac{1}{r^{\gamma-1}}$$

Substituting in equation, we get

$$\eta_{th} = 1 - \frac{1}{\gamma}\left(\frac{1}{r}\right)^{\gamma-1}\left[\frac{\left(\frac{T_3}{T_2}\right)^{\gamma} - 1}{\frac{T_3}{T_2} - 1}\right]$$

Now $\frac{T_3}{T_2} = \frac{V_3}{V_2} = r_c = \text{cut-off ratio}$

$$\eta = 1 - \frac{1}{r^{\gamma-1}}\left[\frac{r_c^{\gamma} - 1}{\gamma(r_c - 1)}\right]$$

It is seen that the expressions are similar except for the term in the parentheses for the Diesel cycle. It can be shown that this term is always greater than unity.

Now $$r_c = \frac{V_3}{V_2} = \frac{V_3}{V_4} / \frac{V_2}{V_1} = \frac{r}{r_e}$$

where r is the compression ratio and r_e is the expansion ratio. Thus, the thermal efficiency of the Diesel cycle can be written as

$$\eta = 1 - \frac{1}{r^{\gamma-1}} \left[\frac{\left(\frac{r}{r_e}\right)^{\gamma} - 1}{\gamma\left(\frac{r}{r_e} - 1\right)} \right]$$

Let r_e = r–Δ since r is greater than r_e. Here, Δ is a small quantity. We therefore have

$$\frac{r}{r_e} = \frac{r}{r-\Delta} = \frac{r}{\left(1-\frac{\Delta}{r}\right)} = \left(1-\frac{\Delta}{r}\right)^{-1}$$

We can expand the last term binomially so that

$$\left(1-\frac{\Delta}{r}\right)^{-1} = 1 + \frac{\Delta}{r} + \frac{\Delta^2}{r^2} + \frac{\Delta^3}{r^3} + \ldots$$

Also $$\left(\frac{r}{r_e}\right)^{\gamma} = \frac{r^{\gamma}}{(r-\Delta)^{\gamma}} = \frac{r^{\gamma}}{r^{\gamma}\left(1-\frac{\Delta}{r}\right)^{\gamma}} = \left(1-\frac{\Delta}{r}\right)^{-\gamma}$$

We can expand the last term binomially so that

$$\left(1-\frac{\Delta}{r}\right)^{-\gamma} = 1 + \gamma\frac{\Delta}{r} + \frac{\gamma(\gamma+1)}{2!}\frac{\Delta^2}{r^2} + \frac{\gamma(\gamma+1)(\gamma+2)}{3!}\frac{\Delta^3}{r^3} + \ldots$$

Substituting in Equation, we get

$$\eta = 1 - \frac{1}{r^{\gamma-1}} \left[\frac{\frac{\Delta}{r} + \frac{(\gamma+1)}{2!}\frac{\Delta^2}{r^2} + \frac{(\gamma+1)(\gamma+2)}{3!}\frac{\Delta^3}{r^3} + \ldots}{\frac{\Delta}{r} + \frac{\Delta^2}{r^2} + \frac{\Delta^3}{r^3} + \ldots} \right]$$

Since the coefficients of $\frac{\Delta}{r}, \frac{\Delta^2}{r^{\gamma}}, \frac{\Delta^3}{r^3},$

etc

are greater than unity, the quantity in the brackets in Equation will be greater than unity. Hence, for the Diesel cycle, we subtract $\frac{1}{r^{\gamma-1}}$ times a quantity greater than unity from one, hence for the same r, the Otto cycle efficiency is greater than that for a Diesel cycle. If $\frac{\Delta}{r}$ is small, the square, cube, etc of this quantity becomes progressively smaller, so the thermal efficiency of the Diesel cycle will tend towards that of the Otto cycle. From the foregoing we can see the importance of cutting off the fuel supply early in the forward stroke, a condition which, because of the short time available and the high pressures involved, introduces practical difficulties with high speed engines and necessitates very rigid fuel injection gear.

In practice, the diesel engine shows a better efficiency than the Otto cycle engine because the compression of air alone in the former allows a greater compression ratio to be employed. With a mixture of fuel and air, as in practical Otto cycle engines, the maximum temperature developed by compression must not exceed the self ignition temperature of the mixture; hence a definite limit is imposed on the maximum value of the compression ratio.

Thus Otto cycle engines have compression ratios in the range of 7 to 12 while diesel cycle engines have compression ratios in the range of 16 to 22.

We can obtain a value of r_c for a Diesel cycle in terms of Q' as follows:

$$r_c = \frac{Q'}{c_p T_1 r^{\gamma-1}} + 1$$

We can substitute the value of ç from Equation in Equation , reproduced below and obtain the value of mep/p_1 for the Diesel cycle.

$$\frac{mep}{p_1} = \eta \frac{Q'}{c_v T_1} \frac{1}{\left[1 - \frac{1}{r}\right][\gamma - 1]}$$

In terms of the cut-off ratio, we can obtain another expression for mep/p_1 as follows:

$$\frac{mep}{p_1} = \frac{\gamma r^{\gamma}(r_c - 1) - r(r_c^{\gamma} - 1)}{(r-1)(\gamma-1)}$$

For the Diesel cycle, the expression for mep/p_3 is as follows:

$$\frac{mep}{p_3} = \frac{mep}{p_1}\left(\frac{1}{r^{y}}\right)$$

Modern high speed diesel engines do not follow the Diesel cycle. The process of heat addition is partly at constant volume and partly at constant pressure. This brings us to the dual cycle.

Dual Cycle

An important characteristic of real cycles is the ratio of the mean effective pressure to the maximum pressure, since the mean effective pressure represents the useful (average) pressure acting on the piston while the maximum pressure represents the pressure which chiefly affects the strength required of the engine structure. In the constant-volume cycle. it is seen that the quantity mep/p_3 falls off rapidly as the compression ratio increases, which means that for a given mean effective pressure the maximum pressure rises rapidly as the compression ratio increases.

For example, for a mean effective pressure of 7 bar and Q'/c_vT_1 of 12, the maximum pressure at a compression ratio of 5 is 28 bar whereas at a compression ratio of 10, it rises to about 52 bar. Real cycles follow the same trend and it becomes a practical necessity to limit the maximum pressure when high compression ratios are used, as in diesel engines. This also indicates that diesel engines will have to be stronger (and hence heavier) because it has to withstand higher peak pressures. Constant pressure heat addition achieves rather low peak pressures unless the compression ratio is quite high. In a real diesel engine, in order that combustion takes place at constant pressure, fuel has to be injected very late in the compression stroke (practically at the top dead center). But in order to increase the efficiency of the cycle, the fuel supply must be cut off early in the expansion stroke, both to give sufficient time for the fuel to burn and thereby increase combustion efficiency and reduce after burning but also reduce emissions. Such situations can be achieved if the engine was a slow speed type so that the piston would move sufficiently slowly for combustion to take place despite the late injection of the fuel.

For modern high speed compression ignition engines it is not possible to achieve constant pressure combustion. Fuel is injected somewhat earlier in the compression stroke and has to go through the various stages of combustion. Thus it is seen that combustion is nearly at constant volume (like in a spark ignition engine). But the peak pressure is limited because of strength considerations so the rest of the heat addition is believed to take place at constant pressure in a cycle.

This has led to the formulation of the dual combustion cycle. In this cycle, for high compression ratios, the peak pressure is not allowed to increase beyond a certain limit and to account for the total addition, the rest of the heat is assumed to be added at constant pressure. Hence the name *limited pressure cycle.*

The cycle is the equivalent air cycle for reciprocating high speed compression ignition engines. The P-V and T-s diagrams are shown in Figs. In the cycle, compression and expansion processes are isentropic; heat addition is partly at constant volume and partly at constant pressure while heat rejection is at constant volume as in the case of the Otto and Diesel cycles.

The heat supplied, Q_s per unit mass of charge is given by

$$c_{v(}T_3 - T_2) + c_{p(}T_{3'} - T_2)$$

whereas the heat rejected, Q_r per unit mass of charge is given by

$c_{v(}T_4 - T_1)$

and the thermal efficiency is given by

$$\eta_{th} = 1 - \frac{c_v(T_4 - T_1)}{c_v(T_3 - T_2) + c_p(T_{3'} - T_2)}$$

$$= 1 - \left\{ \frac{T_1\left(\frac{T_4}{T_1} - 1\right)}{T_2\left(\frac{T_3}{T_2} - 1\right) + \gamma T_3\left(\frac{T_{3'}}{T_3} - 1\right)} \right\}$$

$$= 1 - \frac{\frac{T_4}{T_1} - 1}{\frac{T_2}{T_1}\left(\frac{T_3}{T_2} - 1\right) + \frac{\gamma T_3}{T_2}\frac{T_2}{T_1}\left(\frac{T_{3'}}{T_3} - 1\right)}$$

From thermodynamics $\frac{T_3}{T_2} = \frac{p_3}{p_2} = r_p$

the explosion or pressure ratio and $\frac{T_{3'}}{T_3} = \frac{V_{3'}}{V_3} = r_c$

the cut-off ratio.

Now, $\frac{T_4}{T_1} = \frac{p_4}{p_1} = \frac{p_4}{p_{3'}}\frac{p_{3'}}{p_3}\frac{p_3}{p_2}\frac{p_2}{p_1}$

Also $\frac{p_4}{p_{3'}} = \left(\frac{V_{3'}}{V_4}\right)^{\gamma} = \left(\frac{V_{3'}}{V_3}\frac{V_3}{V_4}\right)^{\gamma} = \left(r_c\frac{1}{r}\right)^{\gamma}$

And $\frac{p_2}{p_1} = r^{\gamma}$

Thus $\frac{T_4}{T_1} = r_p r_c^{\gamma}$

Also $\frac{T_2}{T_1} = \left(\frac{V_1}{V_2}\right)^{\gamma} = r^{\gamma - 1}$

Therefore, the thermal efficiency of the dual cycle is

$$\eta = 1 - \frac{1}{r^{\gamma - 1}}\left[\frac{r_p r_c^{\gamma} - 1}{(e_p - 1) + \gamma r_p (r_c - 1)}\right]$$

We can substitute the value of ç from Equation in Equation 26 and obtain the value of mep/p_1 for the dual cycle.

In terms of the cut-off ratio and pressure ratio, we can obtain another expression for mep/p_1 as follows:

$$\frac{mep}{p_1}=\frac{\gamma r_p r^{\gamma}(r_c-1)+r^{\gamma}(r_p-1)-r(r_c^{\gamma}-1)}{(r-1)(\gamma-1)}$$

For the dual cycle, the expression for mep/p_3 is as follows:

$$\frac{mep}{p_3}=\frac{mep}{p_1}\left(\frac{p_1}{p_3}\right)$$

Since the dual cycle is also called the limited pressure cycle, the peak pressure, p_3, is usually specified. Since the initial pressure, p_1, is known, the ratio p_3/p_1 is known. We can correlate r_p with this ratio as follows:

$$r_p=\frac{p_3}{p_1}\left(\frac{1}{r^{\gamma}}\right)$$

We can obtain an expression for r_c in terms of Q' and r_p and other known quantities as follows:

$$r_p=\frac{1}{\gamma}\left(\left[\left\{\frac{Q'}{c_v T_1 r^{\gamma-1}}\right\}\frac{1}{r_p}\right]+(\gamma+1)\right)$$

We can also obtain an expression for r_p in terms of Q' and r_c and other known quantities as follows:

$$r_p=\frac{\left[\dfrac{Q'}{c_v T_1 r^{\gamma-1}}+1\right]}{1+\gamma r_c-\gamma}$$

Figure shows a constant volume and a constant pressure cycle, compared with a limited pressure cycle. In a series of air cycles with varying pressure ratio at a given compression ratio and the same Q', the constant volume cycle has the highest efficiency and the constant pressure cycle the lowest efficiency. Figure compares the efficiencies of the three cycles for the same value of $Q'\left(\frac{r}{r-1}\right)$ for the same initial conditions and three values of p_3/p_1 for the dual cycle. It is interesting to note that the air standard efficiency is little affected by compression ratio above a compression ratio of 8 for the limited pressure cycle.

The curves of mep/p_3 versus compression ratio for the same three cycles as above are given in Fig.. It is seen that a considerable increase in this ratio is obtained for a limited pressure cycle as compared to the constant volume or constant pressure cycles.

ENGINE EFFICIENCIES AND TERMS

ENGINE EFFICIENCIES

Engines are a good place to start when looking at improving operating efficiencies of access equipment. One benefit of statutory-driven emissions reductions over the recent years has been the improved fuel economies - more efficient burning means the release of more energy.

However, the forthcoming Stage IIIB/interim Tier 4 requirements are focusing engine designers' minds on the 2011 implementation date for 130 kW to 560 kW engines. One consequence, at least according to Perkins' product marketing manager Allister Dennis, is that all those engines will require turbochargers. These systems will give engines, in among other things, greater power density.

Meanwhile, as long as oil prices stay high, there are likely to be more hybrid drive systems developed and fitted into access equipment. Hybrid drive systems usually include electric motors and additional battery capacity along with diesel engines.

The diesel engine charges the batteries and provides mechanical power, according to load and speed. Many hybrid systems include components that capture the energy otherwise dissipated as heat when braking to charge the electrical system.

Hybrid systems are expensive compared to traditional diesel-only power trains, but persistently high fuel prices have made them more economically attractive.

Deutz, working with engine control system specialist Heinzmann, has announced the development of a hybrid package that it says could reduce fuel consumption by 30%. Its electrical system is rated at 15 kW output, with a peak capacity of up to 30 kW, almost doubling the power of the diesel engine.

A prototype unit has already been fitted to an item of construction equipment, replacing its standard 51 kW Deutz TCD 2011 diesel engine with a 37 kW D 2011 non-turbo-charged engine. That is a 28% reduction in engine power. Deutz says the earliest the hybrid unit is likely to go into series production is 2010.

Improving operating efficiencies is not limited to internal combustion engine technology, as Volvo demonstrates. It is helping to develop a new type of battery, called Effpower. The battery, based on proven lead-acid technology, has doubled power, Volvo claims, while manufacturing costs "can be significantly reduced compared with alternatives."

Improvements also come from better integrating the various components in a drive drain. One example is electronic engine control using data-bus systems, which much improves the communication between engine and transmission. Similar integration is found to match hydraulic pump loads to their power sources, regulating engine speed to the demand from boom and locomotive drives. Other efficiency gains are possible with new transmissions. One example is ZF's AS-Tronic drive with dry clutch, which is more efficient than a torque converter. The transmission has double the number of gears, 12 instead of six. That means that equipment in which it is installed require less fuel, as the engine always runs at its optimum speed. So 15 to 20% fuel savings were realised."

According to ZF, other benefits of the AS Tronic automated manual transmission come from its 'intelligent' shifting feature. It lets the engine run in the most fuel-economical speed range, according to the company.

"Quickly changing gears reduces engine idling, and shorter shifting times means shorter power interruptions. The AS Tronic shifts each gear reliably and correctly. This conserves the entire driveline and increases the life of the components," says the company. Overall, the manual transmission adds considerably to the service life of the clutch, adds the company, as does the advantage of having fewer mechanical parts on the inside of the transmission and no mechanical connections to a control station. Mike Goatley, ZF sales manager for off highway products in the UK, says the trend will continue towards greater communication between all components of the driveline, including the engine, transmission, axles and brakes.

"As a complete driveline supplier, ZF is now working on systems where the entire driveline is operated automatically - transmission shifts, braking functions, differential locks, etc. This allows the driveline to achieve maximum efficiency with no driver input."

Much access equipment have a hydraulic pump that delivers a fixed flow of oil. In many operating conditions, a large part of the pressurised (read 'energy') flow circulates back into the tank. Some of the excess energy converts to heat in the control system, and, if the oil must be kept below 70 degrees, an oil cooler is needed," explains Lars Anderson, Hiab R&D manager, structural mechanics. A solution is to use a variable pump to adjust flows to match the hydraulic cylinder and motor displacements to produce desired boom and traction motion.

IMPROVING IC ENGINE EFFICIENCY

Fuel 100%

Pushing the pistons 35%

Overcoming engine friction and pumping air and fue (typical US driving condition) 20%

Are we stuck with ~20% auto engine efficiency?

What can be done?

- Run the engine fuel-lean, that is, use excess air. It is well known that fuel-lean running improves the efficiency. In the old days, under cruising conditions, the engines always ran lean – about 15% excess air — this was economical. So what happen to change this? The problem is the three-way (CO, UHC, NOx) catalyst used on engine exhausts. This only works if the engine air/fuel ratio (by mass) is stoichiometric (chemically correct). For gasoline this ratio is 14.6:1. The engine computer, acting in concert with the engine air flow sensor, electronic fuel injectors, and exhaust oxygen sensor, maintains the stoichiometric ratio for most of your driving. Only at this ratio can the catalyst both oxidize the CO and UHC (to CO_2 and H_2O) and chemically reduce the NOx (to N_2). (UHC = unburned hydrocarbons.) What humankind needs is a lean-NOx catalyst. Then we could have increased efficiency and continue to be clean!

Also needed are ways to improve lean flammability in gasoline engines. That is, the ability to burn real lean is limited by the fuel. If the gasoline-air mixture is too lean, the flame will not have enough speed to get across the cylinder in the time permitted by the engine RPM the driver wants, or the flame will not even start – the cylinder misfires, and then the catalyst has to oxidize a huge amount of UHC and thus may overheat (which might mean you have to buy a new catalyst).

Background

A first course on thermodynamics may teach the efficiency of the Otto cycle (which is the ideal cycle used to simulate the gasoline spark ignition auto engine). Such a course would derive the following equation for the Otto cycle efficiency:

$$h = 1 - 1/r_v^{g-1}$$

The compression ratio of the engine is r_v. Actually, this is a volume ratio. It is the ratio of the volume in a cylinder when the piston is at the bottom of the cylinder to the volume in the cylinder when the piston is at its top position:

$$r_v = V_{bottom}/V_{top}.$$

Most auto engines have compression ratios in the 9 to 10.5 range. We note: the higher the compression ratio, the higher the efficiency! The g parametre is the ratio of the specific heats, ie, the constant pressure specific heat over the constant volume specific heat. In practical terms, the higher the g, the higher the efficiency.

A gas such as helium or argon, composed only of atoms, has the highest g possible, 1.67. Room air on the other hand, being mainly composed of O_2 and N_2 molecules has a g of 1.4. Fuel vapour hasg less than that of air. The mixture of air and gasoline vapour inducted into the engine has a g of about 1.35. As this mixture is compressed and heated during the compression stroke, its g drops to about 1.33. Upon combustion (when the piston is near its top position), the fuel is oxidized to CO_2 (and some CO) and H_2O, and g drops further. It drops into the 1.20-1.25 range. The overall, effective g for the whole cycle for use in the efficiency equation above is about 1.27.

The rule of thumb is: the greater the complexity of the molecules, the lower the g. The lower limit is 1. Argon and helium atoms only translate, that is, they move along straight paths until they encounter another atom. Room air molecules translate and rotate (about 2 of their axes). Hot air starts to vibrate (as two nuclei connected by a spring). Molecules of fuel vapour have a lot of opportunity to vibrate, even at room temperature. The products of combustion vibrate.

However, only the translation of the molecules PUSHES the piston. The other modes of molecular motion do nothing for pushing the piston. Thus, as gdrops (indicating more vibration of the molecules), h drops. A lean engine (ie, an engine with excess air) has a cooler combustion process and more air relative to fuel than the typical engine with a chemically correct mixture. Thus, its g is higher, and its h is greater.

Plug g = 1.27 into the efficiency equation above, assume r_v = 10, and you get h = 0.46. Multiply this by about 0.75 to account for real cycle effects (such as the time it takes to burn, heat losses to the coolant, and exhaust valves that open before the piston fully reaches bottom position) and you have h = 0.35. This is the efficiency (given above) of using the chemical energy of the fuel to push the pistons. Multiply this by the mechanical efficiency of the engine, which accounts for the mechanical friction in the engine and for the air (and fuel) pumping work that has to be done, and you have the final, or overall efficiency of the engine. Of course the mechanical efficiency varies with driving conditions. The higher the RPM of the engine, the greater the friction loss. The more closed the throttle (ie, the farther your foot is off the pedal), the higher the pumping loss. For typical US driving, the resultant overall efficiency of the engine is about 20%. Note, your pedal is not really a gas pedal, it is an air pedal! Add the tranny and real axle mechanical friction losses (or the transaxle friction losses) and the drain of a few essential accessories, and you arrive at a 15% fuel-to-wheel efficiency for the typical auto driven in the US.

- Higher compression ratio. Here, we are limited by autoignition of the gasoline – knock. That is, if the gasoline engine compression is above about 10.5, unless the octane number of the fuel is high, knocking combustion occurs. This is annoying and if persistent, damage to the engine can occur. Thus, gasoline engines are limited in their efficiency by the inability of the fuel to smoothly burn in high compression ratio engines.

  However, the diesel engine is not subject to this limitation. It runs at high compression ratio. In part, this explains its high efficiency. It also runs lean, and its pumping work is low, further increasing its efficiency over the gasoline engine. Humankind needs quiet, smoke-free, odor-free diesels!

- We need new cycles put into practical use. An example is the Atkinson cycle. This has a smaller compression ratio than expansion ratio. This means T_C is reduced since the burnt gas cool as they expand, making the cycle efficient. We throw away less waste heat via the exhaust.
- Run the engine at optimum conditions, meaning low friction (modest engine speed) and low pumping work (air throttle more open). Try to approach the "pushing-the-pistons" efficiency of 35%. This already is happening in some stationary piston engines – large, slow, piston engines used at pipeline compressor stations, for example. Also, this is an important characteristic of the engines used in the hybrid gasoline-electric vehicles. Let the gasoline engine in the hybrid gasoline-electric power plant only run with good throttle opening and modest RPM. An example of one type of commercially available hybrid engine (a "parallel" type) is found at:

Note the hybrid power plant also recovers some of the kinetic energy of the vehicle, by letting this KE drive an electrical generator (during braking). The electrical energy is stored in the batteries. (Normally, this KE is dissipated as heat in the brakes.) An inverter is used to convert DC electricity from the batteries to AC electricity needed by the electric motor and created by the generator.

The table below compares the "well-to-wheel" efficiencies of several auto power plants. "Fuel Prod" means the energy efficiency of extracting, refining, and transporting the fuel. "Eng" means the "fuel-to-wheel"efficiency of the vehicle. "Gas" means gasoline engine. "FC-HC" means a PEM fuel cell with a gasoline-to-hydrogen reformer on board. (PEM = proton exchange membrane fuel cell, the fuel cell type that has been getting most of the attention for auto and home use.) "FC-MeOH" means a PEM fuel cell with methanol-to-hydrogen reformer on board. (The methanol is produced at a refinery by steam reforming natural gas – thus it is a "fossil fuel".) "Ems" means emissions (CO, UHC, NOx) impact. The ratings are "low" (where we are now for autos), "ultra low", and super low".

CALCULATION OF EFFICIENCY

Efficiency is a very easy concept. The calculations you will be given in your GCSE exam should not give you any problems at all. You just need to be able to calculate percentages. You might be asked to calculate the efficiency of a transformer. What you must do is think about how much energy the transformer draws from the mains and how much of this energy is used in a useful way. ie What percentage of the energy is used usefully? It would be nice to say 100 per cent but it is probably only 80 per cent efficient. Well nothing is perfectly efficient. If you have a mobile phone, you will have noticed that the charger gets warm when you use it to recharge the battery of your mobile. This means that some of the electrical energy has been wasted ie turned into heat.

You can actually do the calculation for yourself if you already have a mobile phone and a charger. I hope that it is as cool as my mobile phone. On the bottom of my charger it says that it uses 21 mAmps at 230 volts. I can now calculate how much power it draws from the mains.

Remember P = VI.

Power = 230 volts x 21 mAmps = 4.83 watts

To get the answer in watts I had to divide by 1000 because I started with mAmps instead of Amps. The charger also says that it has an output of 355 mAmps at 3.7 volts. So I can calculate how much of the power goes into the battery of my mobile phone.

Power = 3.7 volts x 355 mAmps = 1.31 watts

Well that is a bit disappointing. It seems that quite a lot of the energy is being wasted. My mobile is cool, but the charger is not very efficient and does get quite hot when I use it to recharge the phone. So how efficient is it?

Power is the rate of doing work. So every second it is charging the phone, it uses 4.83 joules of energy, but only puts 1.31 joules into the battery. The other 3.52 joules per second is wasted. Here is the calculation of per cent efficiency:

% efficiency = 1.31 x 100/ 4.83 = 27 %

That is not quite the end of the story for my mobile phone. The transformer is only 27 per cent efficient, but there is also a loss of energy in the mobile when it is charging. I know this because the battery also gets hot by the time it is fully charged. Some of the electrical energy is converted into chemical energy in the battery and the rest is wasted as heat.

You can calculate the per cent efficiency of any machine provided that you know how much energy has to be put into it and how much useful energy comes out. Here is the equation:

% efficiency = useful energy produced x 100/ total energy used

METHODS FOR CALCULATING EFFICIENCY

CHP is an efficient and clean approach to generating power and thermal energy from a single fuel source.

CHP is used either to replace or supplement conventional separate heat and power (SHP) (i.e., central station electricity available via the grid and an onsite boiler or heater).

- CHP System Efficiency Defined
- Key Terms Used in Calculating CHP Efficiency

- Calculating Total System Efficiency
- Calculating Effective Electric Efficiency
- Which CHP Efficiency Metric Should You Select?

CHP System Efficiency Defined

Every CHP application involves the recovery of otherwise wasted thermal energy to produce additional power or useful thermal energy. Because CHP is highly efficient, it reduces emissions of traditional air pollutants and carbon dioxide, the leading greenhouse gas associated with global climate change.

Efficiency is a prominent metric used to evaluate CHP performance and compare it to SHP. This identifies and describes the two methodologies most commonly used to determine the efficiency of a CHP system: *total system efficiency and effective electric efficiency*.

The illustration below illustrates the potential efficiency gains of CHP when compared to SHP.

Conventional Generation vs. CHP: Overall Efficiency

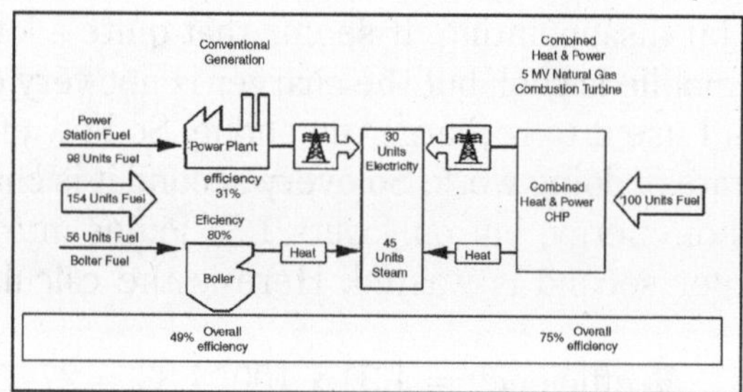

In this example of a typical CHP system, to produce 75 units of useful energy, the conventional generation or separate heat and power systems use 154 units of energy—98 for electricity production and 56 to produce heat—resulting in an overall efficiency of 49 percent. However, the CHP system needs only 100 units of energy to produce the 75 units of useful energy from a single fuel source, resulting in a total system efficiency of 75 percent.

Key Terms Used in Calculating CHP Efficiency

Calculating a CHP system's efficiency requires an understanding of several key terms, described below.

- CHP system. The CHP system includes the unit in which fuel is consumed (*e.g.*, turbine, boiler, engine), the electric generator, and the heat recovery unit that transforms otherwise wasted heat to useable thermal energy.
- Total fuel energy input (Q_{FUEL}). The thermal energy associated with the total fuel input. Total fuel input is the sum of all the fuel used by the CHP system. The total fuel energy input is often determined by multiplying the quantity of fuel consumed by the heating value of the fuel.

Commonly accepted heating values for natural gas, coal, and diesel fuel are:

- 1020 Btu per cubic foot of natural gas
- 10,157 Btu per pound of coal
- 138,000 Btu per gallon of diesel fuel

- Net useful power output (W_E). Net useful power output is the gross power produced by the electric generator minus any parasitic electric losses in other words, the electrical power used to support the CHP system. (An example of a parasitic electric loss is the electricity that may be used to compress the natural gas before the gas can be fired in a turbine.)
- Net useful thermal output (ΣQ_{TH}). Net useful thermal output is equal to the gross useful thermal output of the CHP system minus the thermal input. An example of thermal input is the energy of the condensate return and makeup water fed to a heat recovery steam generator (HRSG). Net useful thermal output represents the otherwise wasted thermal energy that was recovered by the CHP system.

Gross useful thermal output is the thermal output of a CHP system *utilised* by the host facility. The term utilised is important here. Any thermal output that is not used should not be considered. Consider, for example, a CHP system that produces 10,000 pounds of steam per hour, with 90 percent of the steam used for space heating and the remaining 10 percent exhausted in a cooling tower. The energy content of 9,000 pounds of steam per hour is the gross useful thermal output.

Calculating Total System Efficiency

The most commonly used approach to determining a CHP system's efficiency is to calculate *total system efficiency*. Also known as *thermal efficiency*, the total system efficiency (η_o) of a CHP system is the sum of the net useful power output (W_E) and net useful thermal outputs (ΣQ_{TH}) divided by the total fuel input (Q_{FUEL}), as shown below:

$$\eta_o = \frac{W_E + \Sigma Q_{TH}}{Q_{FUEL}}$$

The calculation of total system efficiency is a simple and useful method that evaluates what is produced (*i.e.*, power and thermal output) compared to what is consumed (*i.e.*, fuel). CHP systems with a relatively high net useful thermal output typically correspond to total system efficiencies in the range of 60 to 85 percent.

Note that this metric does not differentiate between the value of the power output and the thermal output; instead, it treats power output and thermal output as additive properties with the same relative value. In reality and in practice, thermal output and power output are not interchangeable because they cannot be converted easily from one to another. However, typical CHP applications have coincident power and thermal demands that must be met. It is reasonable, therefore, to consider the values of power and thermal output from a CHP system to be equal in many situations.

Calculating Effective Electric Efficiency

Effective electric efficiency calculations allow for a direct comparison of CHP to conventional power generation system performance (*e.g.*, electricity produced from central stations, which is how the majority of electricity is produced in the United States).

Effective electric efficiency (ξEE) can be calculated using the equation below, where (W_E) is the net useful power output, (ΣQ_{TH}) is the sum of the net useful thermal outputs, (Q_{FUEL}) is the total fuel input, and " equals the efficiency of the conventional technology that otherwise would be used to produce the useful thermal energy output if the CHP system did not exist:

$$\varepsilon_{EE} = \frac{W_E}{Q_{FUEL} - \Sigma(Q_{TH}/\alpha)}$$

For example, if a CHP system is natural gas fired and produces steam, then a represents the efficiency of a conventional natural gas-fired boiler. Typical a values for boilers are: 0.8 for natural gas-fired boiler, 0.75 for a biomass-fired boiler, and 0.83 for a coal-fired boiler.

The calculation of effective electric efficiency is essentially the CHP net electric output divided by the additional fuel the CHP system consumes over and above what would have been used by conventional systems to produce the thermal output for the site. In other words, this metric measures how effectively the CHP system generates power once the thermal demand of a site has been met.

Typical effective electrical efficiencies for combustion turbine-based CHP systems are in the range of 51 to 69 percent. Typical effective electrical efficiencies for reciprocating engine-based CHP systems are in the range of 69 to 84 percent.

Which CHP Efficiency Metric Should You Select

The selection of an efficiency metric depends on the purpose of calculating CHP efficiency.

- If the objective is to compare CHP system energy efficiency to the efficiency of a site's SHP options, then the total system efficiency metric may be the right choice. Calculation of SHP efficiency is a weighted average (based on a CHP system's net useful power output and net useful thermal output) of the efficiencies of the SHP production components. The separate power production component is typically 33 percent efficient grid power. The separate heat production component is typically a 75- to 85-percent efficient boiler.

- If CHP electrical efficiency is needed for a comparison of CHP to conventional electricity production (*i.e.*, the grid), then the effective electric efficiency metric may be the right choice. Effective electric efficiency accounts for the multiple outputs of CHP and allows for a direct comparison of CHP and conventional electricity production by crediting that portion of the CHP system's fuel input allocated to thermal output.

Both the total system and effective electric efficiencies are valid metrics for evaluating CHP system efficiency. They both consider all the outputs of CHP systems and, when used properly, reflect the inherent advantages of CHP. However, since each metric measures a different performance characteristic, use of the two different metrics for a given CHP system produces different values.

For example, consider a gas turbine CHP system that produces steam for space heating with the following characteristics:

Fuel Input (MMBtu/hr)	41
Electric Output (MW)	3.0
Thermal Output (MMBtu/hr)	17.7

Using the total system efficiency metric, the CHP system efficiency is 68 percent (3.0*3.413+17.7)/41).

Using the effective electric efficiency metric, the CHP system efficiency is 54 percent (3.0*3.413)/(41-(17.7/0.8).

This is not a unique example; a CHP system's total system efficiency and effective electric efficiency often differ by 5 to 15 percent.

NOTE: Many CHP systems are designed to meet a host site's unique power and thermal demand characteristics. As a result, a truly accurate measure of a CHP system's efficiency may require additional information and broader examination beyond what is described.

OTTO

DEFINITION

OTTO is a Stove Top Espresso Maker, combining classic Italian style with unprecedented functionality. Meticulously engineered and crafted, "the little guy" offers inspirational ergonomics and delivers superb coffee. OTTO is made almost completely of stainless steel promising you a lifetime of satisfaction from this modern international design classic.

Recognised for excellence, the 'Little Guy' won the Australian Design Mark at the prestigous 2008 Australian International Design Awards and is currently a finalist in the International Design Excellence Awards in America (IDEA).

The three year development phase has been driven by a love for great coffee. The process of uncompromised care and attention has produced results beyond initial expectations.

The critical issues of extracting great coffee and steaming milk with an exceptional texture have been comprehensively achieved. OTTO comes with an OTTO tamper, the OTTO experience DVD providing a barista training session, two Italian designed latte glasses and a stainless milk jug, packaged in a robust premium travel case.

OTTO CONTROLS

For over 40 years OTTO has been designing and manufacturing a full line of switches and operator controls for demanding applications. Our switch product line includes digital and analog output, sealed and lighted, snap action, rocker, pushbutton, toggle and rotary switches. We provide operator control grips for military and commercial applications including joysticks with Hall Effect technology, J1939, CAN & PWM communication protocols.

OTTO evaluates every customer need by starting with an engineering product review to gain a more thorough understanding of the application and needs. OTTO has the ability and expertise to provide a customer with a recommended switch for their application, the ability to mold the housing as well as the cabling and assembly.

Our integrated manufacturing facilities enable us to design and fabricate an entire solution from the switch to the control grip.

The result is a cost effective design that meets performance and quality objectives for any customer.

Our Engineering and Manufacturing Capabilities Include:

- Electronic Data Interchange
- PRO-E and SDRC 3-D solid modeling work stations
- SLS and SLA compatible outputs for rapid prototyping
- COSMOS Finite Element Analysis
- Mechanical, Electrical and Environmental Testing
- Fabrication assembly areas
- MIL-I-45208 test lab facility with environmental, mechanical, electrical and audio test capabilities
- Computer Controlled Production Line Testing
- ISO 9001-2000 Registered Total Quality Management System
- Certified FAA Repair Station

OTTO is recognized worldwide for superior performance and innovative products. OTTO is a vertically integrated manufacturer with over 200,000 square feet of manufacturing capabilities; providing in house injection molding, stamping, multi axis CNC machining, cable over molding and final assembly work cells.

FEATURES

- Designed for armrest and panel mounting
- Proven contactless analog output Hall Effect technology
- Electronics sealed to IP68S
- Up to 10 million operational cycles in all directions
- Available with a vraiety of grip and switch options
- Redundant sensors available
- Various output configurations

The JHM series Medium Hall Effect Joystick is a full function operator control in a package that will fit in an armrest or on a panel.

It utilizes OTTO's patented Hall Effect technology for unmatched life and reliability. Electronics are sealed and it has an operational life up to 10 million cycles in all directions.

Additional options include CANopen and CAN J1939 versions, multiple analog and digital auxiliary control outputs, redundant sensors and a variety of output configurations, along with a variety of switch options.

ENGINE EFFICIENCIES AND PERFORMANCE

With the use of automotive technology advancements, fuel injection systems have been incorporated with powerful and modern engine assemblies like those you get with a Pontiac ride. By atomising the metered amounts of fuel delivered through the combustion chambers, your engine establishes a cleaner and more efficient fuel burn. While it works with other engine systems to deliver its peak fuel economy status, precision among components and concerned engine systems is very important. Maintaining engine precision is simplest when you pay attention to the heeds of already failing and defective fuel system components particularly on your stock Pontiac fuel injector. As the part that actually sprays the fuel while maintaining precision flow of fuel, it needs to be kept responsive to actuations and effective like new to enjoy smooth combustion mechanisms. Otherwise, you might face poor fuel economy and fuel mileage with severe engine damages. For preventive maintenance, make it a habit to periodically pop the hood and check the actual working state of the part. This enables you to precisely determine the best time to have it repaired or replaced which helps eliminate the risks of engine failure.

Due to stricter emission requirements, modern engine assemblies have undergone series of modifications which gave birth to the use of fuel injection systems. Aiming to have your engine yield excellent air-fuel ratio, Pontiac fuel injector constructions have undergone the same extent of product development research in realising the combustion efficiency its great functionality contributes. As the fuel pump forces the fuel through its designated fuel lines, metered amounts of fuel passes through the injector to be delivered in atomised form. To do that, the part makes use of compression and release mechanisms so its nozzle and valve efficiently regulates the precision flow of fuel which solely relies on the vacuum created by pressure build-up. While engines initially use single point or central fuel injection systems, fuel injection system applications have gradually evolved into multi-port fuel injection systems. While the system is developed to spray fuel right at the intake valve, it surely provides more accurate fuel metering and delivery which is crucial to establishing quicker engine responses.

The incorporation of computer assisted engine operations made it easier to upgrade critical factory settings for engine combustion processes. This way, the engine control unit monitors the efficiency and output of fuel injections systems with greater accuracy and efficiency. With the use of practical and highly sensitive electronic sensors, timely adjustments to sustaining smooth operations while meeting the power and torque demands of specific driving conditions has never been easier and more convenient. This automotive engineering advancement gave birth to electronically controlled Pontiac fuel injector constructions. While the engine control unit effectively keeps track of the pulse width prior to actual combustion operations, you can continually enjoy outstanding fuel efficiency as well as performance gains. You do not need to be a skilled mechanic to determine whether your stock fuel injectors start fouling up. Sluggish engine performance is a result to greatly diminished amount of generated engine power. Noticing increased emission is a result of significantly depreciated combustion efficiencies. Grinding noises right under the hood means that your engine struggles to maintain engine precision while sustaining smooth engine operations. At the first sign of Pontiac fuel injector failure, consult a qualified service technician to stop the progress of potentially serious engine problems due to imprecision. Professional fuel injection system diagnostics will be followed by necessary adjustments and repairs. In case your stock fuel injector fuel injectors go beyond repair, finding and installing equally dependable replacements is advised.

IMPROVING IC ENGINE EFFICIENCY

Fuel 100%

Pushing the pistons 35%

Overcoming engine friction and pumping air and fue (typical US driving condition) 20%

Are we stuck with ~20% auto engine efficiency?

What can be done?

- Run the engine fuel-lean, that is, use excess air. It is well known that fuel-lean running improves the efficiency. In the old days, under cruising conditions, the engines always ran lean – about 15% excess air — this was economical. So what happen to change this? The problem is the three-way (CO, UHC, NOx) catalyst used on engine exhausts. This only works if the engine air/fuel ratio (by mass) is stoichiometric (chemically correct). For gasoline this ratio is 14.6:1. The engine computer, acting in concert with the engine air flow sensor, electronic fuel injectors, and exhaust oxygen sensor, maintains the stoichiometric ratio for most of your driving. Only at this ratio can the catalyst both oxidize the CO and UHC (to CO_2 and H_2O) and chemically reduce the NOx (to N_2). (UHC = unburned hydrocarbons.) What humankind needs is a lean-NOx catalyst. Then we could have increased efficiency and continue to be clean!

Also needed are ways to improve lean flammability in gasoline engines. That is, the ability to burn real lean is limited by the fuel. If the gasoline-air mixture is too lean, the flame will not have enough speed to get across the cylinder in the time permitted by the engine RPM the driver wants, or the flame will not even start – the cylinder misfires, and then the catalyst has to oxidize a huge amount of UHC and thus may overheat (which might mean you have to buy a new catalyst).

Background

A first course on thermodynamics may teach the efficiency of the Otto cycle (which is the ideal cycle used to simulate the gasoline spark ignition auto engine). Such a course would derive the following equation for the Otto cycle efficiency:

$$h = 1 - 1/r_v^{g-1}$$

The compression ratio of the engine is r_v. Actually, this is a volume ratio. It is the ratio of the volume in a cylinder when the piston is at the bottom of the cylinder to the volume in the cylinder when the piston is at its top position:

$$r_v = V_{bottom}/V_{top}.$$

Most auto engines have compression ratios in the 9 to 10.5 range. We note: the higher the compression ratio, the higher the efficiency! The g parametre is the ratio of the specific heats, ie, the constant pressure specific heat over the constant volume specific heat. In practical terms, the higher the g, the higher the efficiency.

A gas such as helium or argon, composed only of atoms, has the highest g possible, 1.67. Room air on the other hand, being mainly composed of O_2 and N_2 molecules has a g of 1.4. Fuel vapour hasg less than that of air. The mixture of air and gasoline vapour inducted into the engine has a g of about 1.35. As this mixture is compressed and heated during the compression stroke, its g drops to about 1.33. Upon combustion (when the piston is near its top position), the fuel is oxidized to CO_2 (and some CO) and H_2O, and g drops further. It drops into the 1.20-1.25 range. The overall, effective g for the whole cycle for use in the efficiency equation above is about 1.27.

The rule of thumb is: the greater the complexity of the molecules, the lower the g. The lower limit is 1. Argon and helium atoms only translate, that is, they move along straight paths until they encounter another atom. Room air molecules translate and rotate (about 2 of their axes). Hot air starts to vibrate (as two nuclei connected by a spring). Molecules of fuel vapour have a lot of opportunity to vibrate, even at room temperature. The products of combustion vibrate.

However, only the translation of the molecules PUSHES the piston. The other modes of molecular motion do nothing for pushing the piston. Thus, as gdrops (indicating more vibration of the molecules), h drops. A lean engine (ie, an engine with excess air) has a cooler combustion process and more air relative to fuel than the typical engine with a chemically correct mixture. Thus, its g is higher, and its h is greater.

Plug g = 1.27 into the efficiency equation above, assume $r_v = 10$, and you get h = 0.46. Multiply this by about 0.75 to account for real cycle effects (such as the time it takes to burn, heat losses to the coolant, and exhaust valves that open before the piston fully reaches bottom position) and you have h = 0.35. This is the efficiency (given above) of using the chemical energy of the fuel to push the pistons. Multiply this by the mechanical efficiency of the engine, which accounts for the mechanical friction in the engine and for the air (and fuel) pumping work that has to be done, and you have the final, or overall efficiency of the engine. Of course the mechanical efficiency varies with driving conditions. The higher the RPM of the engine, the greater the friction loss. The more closed the throttle (ie, the farther your foot is off the pedal), the higher the pumping loss. For typical US driving, the resultant overall efficiency of the engine is about 20%. Note, your pedal is not really a gas pedal, it is an air pedal! Add the tranny and real axle mechanical friction losses (or the transaxle friction losses) and the drain of a few essential accessories, and you arrive at a 15% fuel-to-wheel efficiency for the typical auto driven in the US.

- Higher compression ratio. Here, we are limited by autoignition of the gasoline – knock. That is, if the gasoline engine compression is above about 10.5, unless the octane number of the fuel is high, knocking combustion occurs. This is annoying and if persistent, damage to the engine can occur. Thus, gasoline engines are limited in their efficiency by the inability of the fuel to smoothly burn in high compression ratio engines.

  However, the diesel engine is not subject to this limitation. It runs at high compression ratio. In part, this explains its high efficiency. It also runs lean, and its pumping work is low, further increasing its efficiency over the gasoline engine. Humankind needs quiet, smoke-free, odor-free diesels!

- We need new cycles put into practical use. An example is the Atkinson cycle. This has a smaller compression ratio than expansion ratio. This means T_C is reduced since the burnt gas cool as they expand, making the cycle efficient. We throw away less waste heat via the exhaust.
- Run the engine at optimum conditions, meaning low friction (modest engine speed) and low pumping work (air throttle more open). Try to approach the "pushing-the-pistons" efficiency of 35%. This already is happening in some stationary piston engines – large, slow, piston engines used at pipeline compressor stations, for example. Also, this is an important characteristic of the engines used in the hybrid gasoline-electric vehicles. Let the gasoline engine in the hybrid gasoline-electric power plant only run with good throttle opening and modest RPM. An example of one type of commercially available hybrid engine (a "parallel" type) is found at:

Note the hybrid power plant also recovers some of the kinetic energy of the vehicle, by letting this KE drive an electrical generator (during braking). The electrical energy is stored in the batteries. (Normally, this KE is dissipated as heat in the brakes.) An inverter is used to convert DC electricity from the batteries to AC electricity needed by the electric motor and created by the generator.

The table below compares the "well-to-wheel" efficiencies of several auto power plants. "Fuel Prod" means the energy efficiency of extracting, refining, and transporting the fuel. "Eng" means the "fuel-to-wheel"efficiency of the vehicle. "Gas" means gasoline engine. "FC-HC" means a PEM fuel cell with a gasoline-to-hydrogen reformer on board. (PEM = proton exchange membrane fuel cell, the fuel cell type that has been getting most of the attention for auto and home use.) "FC-MeOH" means a PEM fuel cell with methanol-to-hydrogen reformer on board. (The methanol is produced at a refinery by steam reforming natural gas – thus it is a "fossil fuel".) "Ems" means emissions (CO, UHC, NOx) impact. The ratings are "low" (where we are now for autos), "ultra low", and super low".

8

Steam Engines and Boilers

STEAM ENGINES

Steam locomotives were powered by steam engines, and deserve to be remembered because they swept the world through the Industrial Revolution of the 18th and 19th centuries. Steam engines rank with cars, aeroplanes, telephones, radio, and television among the greatest inventions of all time. They are marvels of machinery and excellent examples of engineering, but under all that smoke and steam.

It takes energy to do absolutely anything you can think of—to ride on a skateboard, to fly on an aeroplane, to walk to the shops, or to drive a car down the street. Most of the energy we use for transportation today comes from oil, but that wasn't always the case. A few decades ago, coal was the world's favourite fuel and it powered everything from trains and ships to the ill-fated steam planes invented by American scientist Samuel P. Langley, an early rival of the Wright brothers. What was so special about coal? There's lots of it inside Earth, so it was relatively inexpensive and widely available.

Coal is an organic chemical, which means it's based on the element carbon. Coal forms over millions of years when the remains of dead plants get buried under rocks, squeezed by pressure, and cooked by Earth's internal heat. Lumps of coal are really lumps of energy. The carbon inside them is locked to atoms of hydrogen and oxygen by joints called chemical bonds. When we burn coal on a fire, the bonds break apart and the energy is released in the form of heat.

A steam engine is a machine that burns coal to release the heat energy it contains. It's a bit like a giant kettle sitting on top of a coal fire. The heat from the fire boils the water in the kettle and turns it into steam. But instead of blowing off uselessly into the air, like the steam from a kettle, the steam is captured and used to power a machine. Let's find out how!

How a steam engine works

Crudely speaking, there are four different parts in a steam engine:

1. A fire where the coal burns.

2. A boiler full of water that the fire heats up to make steam.
3. A cylinder and piston, rather like a bicycle pump but much bigger. Steam from the boiler is piped into the cylinder, causing the piston to move first one way then the other. This in and out movement (which is also known as "reciprocating") is used to drive...
4. A machine attached to the piston. That could be anything from a water pump to a factory machine... or even a giant steam locomotive running up and down a railroad.

That's a very simplified description, of course. In reality, there are hundreds or perhaps even thousands of parts in even the smallest locomotive. It's easiest to see how everything works on our little side-on steam locomotive. Inside the locomotive cab, you load coal into the firebox, which is quite literally a metal box containing a roaring coal fire. The fire heats up the boiler—the "giant kettle" inside the locomotive. The boiler in a steam locomotive doesn't look much like a kettle you'd use to make a cup of tea, but it works the same way, producing steam under high pressure. The boiler is a big tank of water with dozens of thin metal tubes running through it (for simplicity, we show only one here, coloured orange).

The tubes run from the firebox to the chimney, carrying the heat and the smoke of the fire with them (shown as red dots inside the tube).

This arrangement of boiler tubes, as they are called, means the engine's fire can heat the water in the boiler tank much faster, so it produces steam more quickly and efficiently. The water that makes the steam either comes from tanks mounted on the side of the locomotive or from a separate wagon called a tender, pulled behind the locomotive. (The tender also carries the locomotive's supply of coal.) The steam generated in the boiler flows down into a cylinder just ahead of the wheels, pushing a tight-fitting plunger, the piston, back and forth. A little mechanical gate in the cylinder, known as an inlet valve (shown in orange) lets the steam in. The piston is connected to one or more of the locomotive's wheels through a kind of arm-elbow-shoulder joint called a crankshaft and connecting rod. As the piston pushes, the crankshaft and connecting rod turn the locomotive's wheels and power the train along. When the piston has reached the end of the cylinder, it can push no further. The train's momentum (tendency to keep moving) carries the crankshaft onwards, pushing the piston back into the cylinder the way it came. The steam inlet valve closes. An outlet valve opens and the piston pushes the steam back through the cylinder and out up the locomotive's chimney. The intermittent chuff-chuff noise that a steam engine makes, and its intermittent puffs of smoke, happen when the piston back and forth in the cylinder.

There's a cylinder on each side of the locomotive and the two cylinders fire slightly out of step with one another to ensure there's always some power pushing the engine along.

Types of Steam Engine

This is called arotary steam engine, because the piston's job is to make a wheel *rotate*. The earliest steam engines worked in an entirely different way. Instead of turning a wheel, the piston pushed a beam up and down in a simple back-and-forth or reciprocatingmotion. Reciprocating steam engines were used to pump water out of flooded coal mines in the early 18th century.

This is called a single-acting steam engine and it's quite an inefficient design because the piston is being powered only half the time. A much better (though slightly more complex) design uses extra steam pipes and valves to make steam push the piston first one way and then the other. This is called a double-acting (or counterflow) steam engine. It's much more powerful because steam is driving the piston all the time.

The first steam engines were very large and inefficient, which means it took huge amounts of coal to get them to do anything. Later engines produced steam at much higher pressure: the steam was produced in a smaller, much stronger boiler so it squeezed out with more force and blew the piston harder. The extra force of high-pressuresteam engines allowed engineers to make them lighter and more compact, and it was this that paved the way for steam locomotives, steam ships, and steam cars.

Coal was a cheap and abundant fuel during the early Industrial Revolution, but the invention of the gasoline (petrol) engine in the mid-19th century heralded a new era: during the 20th century, oil overtook coal as the world's favourite fuel. Steam engines are extremely inefficient, wasting around 80-90 per cent of all the energy they produce from coal. That means they have to burn enormous amounts of coal to produce useful amounts of power.

A steam engine is so inefficient because the fire that burns the coal is totally separate (and often some distance from) the cylinder that turns the heat energy in the steam into mechanical energy that powers the machine. This design is called an external combustion engine because the fire and boiler are outside the cylinder. It's inefficient because energy is wasted as the heat and steam travel from the fire, via the boiler, to the cylinder. Gasoline- and diesel-powered engines are based on a totally different design called an internal combustion engine. The gasoline or diesel fuel is burned inside the cylinder, not outside it, and this makes internal combustion engines considerably more efficient. Oil has many other advantages too: it's cleaner than coal, makes less air pollution, and is much easier to transport in pipes.

That's largely why steam locomotives disappeared from our railroads—diesel locomotives were altogether more convenient. It takes hours to fire up a steam engine before you can use it; you can get a diesel engine running in less than a minute. Steam engines disappeared from factories when electricity became a more convenient way of powering buildings. Who wants to load coal into a factory every day when they can just flick on switches to make things work? But things are not quite what they seem. Steam and coal never did disappear—not exactly. Where does the electricity we use come from? Well, a great deal of it still comes from coal, burned in power plants miles away from our homes and factories. Inside a coal-fired power plant, giant, efficient, steam engines burn coal to make steam, which drives windmill-like devices called steam turbines.

GENERATION OF STEAM

STEAM

Water can exist in the form of solid, liquid and gas as ice, water and steam respectively. If heat energy is added to water, its temperature rises until a value is reached at which the water can no longer exit as a liquid. We call this the "saturation" point and with any further addition of energy, some of the water will boil off as steam. This evapouration requires relatively large amounts of energy, and while it is being added, the water and the steam released are both at the same temperature. Equally, if we can encourage the steam to release the energy that was added to evaporation it, then the steam will condense and water at the same temperature will be formed.

Why use Steam

Steam is produced by evaporation of water, which is a relatively cheap and plentiful commodity in most parts of the world. Its temperature can be adjusted very accurately by the control of its pressure, using simple valves; it carries relatively large amounts of energy in a small mass, and when it is encouraged to condensate back to water, high rates of energy flow (into the material being heated) are obtained, so that the heat using plant does not have to be unduly large.

Thus steam is most economical,flexible and versatile tool for industry wherever heating is required.

Liquid Enthalpy

Liquid enthalpy is the "Enthalpy" (heat energy) in the water when it has been raised to its boiling point to produce steam, and is measured in kJ/kg, its symbol is h_f. (once known as "Sensible Heat")

Enthalpy of Evaporation

The Enthalpy of evaporation is the heat energy to be added to the water (when it has been raised to its boiling point) in order to change it into steam. There is no change in temperature, the steam produced is at the same temperature as the water from which it is produced, but the heat energy added to the water changes its state from water into steam at the same temperature.

When the steam condenses back into water, it gives up its enthalpy of evaporation, which it had acquired on changing from water to steam. The enthalpy of evaporation is measured in kJ/kg its symbol is h_{fg}. Enthalpy of evaporation is also known as latent heat.

The temperature at which water boils increases as the pressure increases. From this it is evident that as the steam pressure increases, the usable heat energy in the steam (enthalpy of evaporation) which is given up when the steam condenses, actually decreases. The sum of the two enthalpies is known as the enthalpy of saturated steam. This enthalpy is the total heat energy, which is stored in the steam.

Table. Extract from the Steam

Pressure(Bar)	Temperatu re °C	Enthalpy in kJ/kg			Volume (m^3/kg)
		Wate r (h_f)	Evaporation (h_{fg})	Steam(h_g)	
0	100	419	2257	2676	1.673
1	120	506	2201	2707	0.881
2	134	562	2163	2725	0.603
3	144	605	2133	2738	0.461
4	152	671	2108	2749	0.374
5	159	641	2086	2757	0.315
6	165	697	2066	2763	0.272
7	170	721	2048	2769	0.240

STEAM VOLUME

If 1 kg (mass) of water (which is 1 litre, by volume) is all converted into steam, the result will be exactly 1kg (mass) of steam. However, the volume occupied by a given mass depends upon its pressure. At atmospheric pressure 1kg of steam occupies nearly 1.673 cubic metres(m3). At a pressure of 1bar abs, that same 1 kg of steam will only occupy 0.1943 m3. Thus, steam should always be generated and distributed at rated boiler pressure and used at possible low pressure.The volume of 1kg of steam at any given pressure is termed its Specific Volume (symbol Vg).

Steam Quality

In practice, steam often carries tiny droplets of water with it and cannot be described as dry saturated steam. Nevertheless, we find that it is usually important that the steam used for process or heating is as dry as possible.

Steam quality is described by it's "dryness fraction" - the proportion of completely dry steam present in the steam being considered. The steam becomes "wet" if water droplets in suspension are present in the steam space, carrying no specific enthalpy of evaporation. "Wet steam" has a heat content substantially lower than that of dry saturated steam at the same pressure. The small droplets of water in wet steam have weight but occupy negligible space. The volume of wet steam is, therefore, less than that of dry saturated steam.

Volume of Wet Steam = Volume of Dry Saturated Steam * Dryness Fraction

Dryness fraction of the steam depends upon the steam boiler design and capacity. For example, Coil type low capacity boilers produce about 30 to 50% wet steam which is not desired in steam heating applications.

Superheated Steam

As long as water is present, the temperature of saturated steam will correspond to the figure indicated for that pressure in the Steam Tables. However, if heat transfer continues after all the water has been evaporated, the steam temperature will again rise. The steam is then called "superheated", and this "superheated steam" can be at any temperature above that of saturated steam at the corresponding pressure. Superheated Steam is totally dry and follows the gas laws.

Saturated steam will condense very readily on any surface which is at a lower temperature and gives up the enthalpy of evaporation which, as we have seen, is the greater proportion of its energy content. On the other hand, when superheated steam gives up some of its enthalpy, it does so by virtue of a fall in temperature.

No condensation will occur until the saturation temperature has been reached, and it is found that the rate at which we can get energy to flow from superheated steam is often less than we can achieve with saturated steam. Even though the superheated steam is at a higher temperature, superheated steam because of its properties, is the natural first choice for power steam requirements, whilst saturated steam is ideal for process and heating applications.

Chemical Energy

The chemical energy, which is contained in coal, gas or other boiler fuel, is converted into heat energy when the fuel is burned. That heat energy is transmitted through the wall of the boiler furnace to the water.

The temperature of the water is raised by this addition of heat energy until saturation point is reached – it boils. The heat energy which has been added and which has had the effect of raising the temperature of the water is known as the liquid enthalpy.

At that point of boiling, the water is termed Saturated Water. The water in our boiler is now at saturation (boiling) point at 100 °C. Heat transfer is still taking place between the furnace walls and the water. The additional enthalpy produced by this heat transfer does not increase the temperature of the water. It evaporates the water, which changes its state into steam. The enthalpy that produces this change of state without change of temperature is known as the enthalpy of evaporation.

BOILER

The job of a boiler is to supply good quality dry steam at the correct pressure at the right time.

Boilers and the associated fire equipment should be designed for efficient operation. They should also be properly sized. A boiler, which has to cope with a peak load above its maximum continuous rating, will operate at reduced efficiency. Pressure may drop and the resultant priming and carry-over will mean that the boiler is unable to do its job of providing good quality steam.

If a boiler has to work at a small per centage of its rating, radiation losses become significant and, again there is a drop in overall efficiency. Clearly, it is not easy to match boiler plant to what is normally a variable steam load. Two or more boilers are more flexible than a single unit which explains the common arrangement of a large boiler for the winter load with a smaller boiler for the summer load.

Boiler Losses

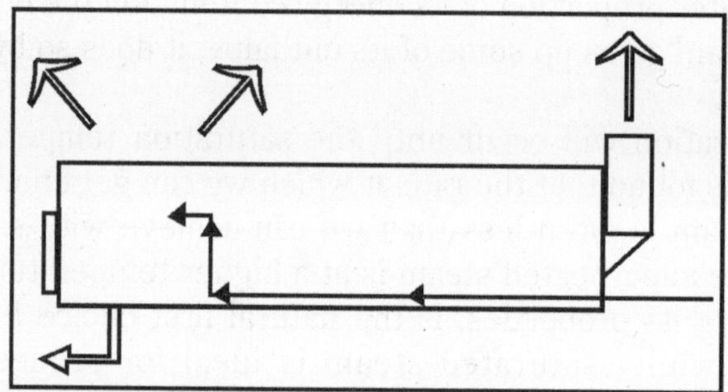

There are four different types of boiler losses:

- Radiation losses
- Flue gas losses (stack losses)
- Blowdown losses
- Losses due to low rate of condensate return from the production plant

The radiation losses are determined through the thermal insulation of the boiler. It should be tried to return as much condensate as possible back to the boiler, so the used heat energy in the boiler can be minimized.

The flue gas and the blowdown losses should be discussed more in detail.

Flue Gas Losses

To achieve a high thermal efficiency, thereby minimizing fuel costs, the amount of combustion air should be limited to that necessary to achieve complete combustion of the fuel.

Excess air is the extra air supplied to the burner beyond the air required for complete or stoichometric combustion. If the supplied air is less, not only will this result in a smoking stack (black smoke indicates the presence of unburned fuel i.e. combustibles in the flue gas), but it will significantly reduce the energy released per unit of fuel. If a burner is operated with a deficiency of air, carbon monoxide and hydrogen will appear in the products of combustion. These combustibles are fuel, and anything in excess of a few hundred parts per million in the flue gas indicates inefficient burner operation.

Too little excess air is inefficient because it permits unburned fuel, in the form of combustibles, to escape up the stack. But too much excess air is also inefficient because it enters the burner at ambient temperature and leaves the stack hot, thus stealing useful heat from the process.

Getting the right mix of fuel and air in your burner is the most critical and difficult step in achieving boiler efficiency. The Oxymiser Combustion Analyser makes sure you always get the right mix of air and fuel.

Blowdown Losses

Boiler blowdown is a very essential function which – if not carried out in the most energy efficient manner – can be an unnecessary source of loss.

Raw water will contain impurities. As the boiler water is evaporated into steam and replaced by make up water, the concentration of the solids in the boiler water will obviously increase. If they are allowed to increase much beyond the recommended level then the functioning of the boiler will be affected. "Foaming" will take place within the boiler resulting in carry-over of water into the steam distribution system, and in the worst situation, malfunctioning of the boiler water level controls can occur.

The original method of "blowing down" a boiler in order to maintain the total dissolved solids (TDS) at an acceptable level was to manually operate a blowdown valve fitted to a discharge pipe at the lowest point of the boiler shell, at the same time hoping that the frequency of operation of the blowdown valve and the length of time for which it is open is sufficient for the needs of the boiler.

Best engineering practice now in the furtherance of energy efficient plant, is to apply the use of an Automatic Blowdown control system (ABCO) which is not necessarily high in capital expenses compared with the payback period in terms of energy saving.

The losses from blowdown may be also be minimized by using the heat in the blowdown water. A certain per centage of the blowdown water will flash off into steam when it enters a region of pressure lower than that which existed within the boiler. This flash steam can be recovered and used by the application of a "Blowdown Heat Recovery Unit" which consists of Blowdown Flash Vessel and Heat Exchanger system to recover the flash steam and use the heat from blowdown water to preheat the Boiler Feed Water. Flash steam is sent to feed water tank for direct heating of water.

The EffiMax 2000 package provides a complete monitoring and data acquisition solution for your boiler performance. EffiMax calculates the efficiency of the boiler based on indirect efficiency computation and computes individually the total amount of losses like stack loss, enthalpy loss, radiation loss and blowdown loss in your boiler. Using the data generated on the system losses, on-line suggestions can then be used to fine tune the system to generate more steam with the lesser quantity of fuel.

Water for the Boiler

The boiler feed tank is the heart of any steam system. It provides a reservoir of returned condensate and fresh make-up water with which the boiler feed pump can replenish the boilers.

The feed tank must be properly sized and allowance made for fluctuations and possible interruptions in supply; it is normal to hold enough water to provide one hour of steam at maximum rating. However, there should be enough free space to cope with the relatively massive return at start-up. Significant quantities of condensate can be lost if this is not provided.

In order to prevent corrosion in boilers and ancillary equipment it is necessary to eliminate both dissolved oxygen and dissolved carbon dioxide from the boiler feed water. Oxygen is the main cause of corrosion and the presence of carbon dioxide in the water presents a pH value of something less than neutral (pH 7) causing the water to be acidic. The ideal pH value for boiler feed water is around pH 9. A higher pH value can cause what is called "caustic embrittlement" in the boiler, particularly a boiler with riveted joints. A Feed Water Tank Systemfrom Spirax provides complete solution to manage the boiler feed water for any capacity boilers.

A Deaerator Head ensures vigorous mixing of steam and feedwater to reduce dissolved oxygen content. This action reduces the need for oxygen scavenging chemicals to a minimum.

EFFICIENCY IN STEAM GENERATION

Every process needs to be optimised for energy efficiency The process of Generating energy is no exception. Efficiency of energy generation is a key factor in the overall profitability of any process plant. Spirax Marshall has a range of unique products, systems and software to help you keep track of exactly how much energy you consume and where it goes.

Our Steam Metreing and Effimax range of products, help you keep an accurate track on parametres that show efficiency of energy generation. Our automatic blowdown control systems ensure a fine control on activities like boiler water blowdown, which is a potential area for energy leakage. Boiler heat recovery units, flash vessels, steam injectors and the Effipro range of products are some of the solutions designed to recover every bit of energy that might be lost in the boilerhouse itself.

INTERNAL ENERGY AND ENTROPY OF STEAM

INTERNAL ENERGY

Internal energy is defined as the energy associated with the random, disordered motion of molecules. It is separated in scale from the macroscopic ordered energy associated with moving objects; it refers to the invisible microscopic energy on the atomic and molecular scale. For example, a room temperature glass of water sitting on a table has no apparent energy, either potential or kinetic . But on the microscopic scale it is a seething mass of high speed molecules traveling at hundreds of metres per second. If the water were tossed across the room, this microscopic energy would not necessarily be changed when we superimpose an ordered large scale motion on the water as a whole. U is the most common symbol used for internal energy.

WORK - W, HEAT - Q, AND INTERNAL ENERGY – U

WORK *W*

Useful Energy Transfered across the System's Boundaries, capable of producing Macroscopic-Mechanical Motion of a the system's Centre-of-Mass.

W = Work done by (or on) one system on another system

ENERGY FLOW			
OUT:	$W > 0$	*System Does External Work*	*Sys --> Work*
INTO:	$W < 0$	*Work Done on the System*	*Work --> Sys*

Work done by a Gas : $W = \int_0^f PdV$

W=0 Constant Volume Process

$W = P\Delta V$ Constant Pressure Process, $W = PV\ln\frac{V_f}{V_0}$

Constant Temperature Process $W = \frac{P_0V_0 - P_fV_f}{\gamma - 1}$

Adiabatic Process Q = 0

HEAT Q

Energy Transfer across the System's Boundaries that cannot produce Macroscopic-Mechanical Motion of the system's Centre-of-Mass. Energy Transfer at the Molecular Level .

Q = Microscopic Energy flow into (or out of) the System,

ENERGY FLOW			
INTO:	$Q > 0$	*System Absorbs Heat*	*Heat --> Sys*
OUT:	$Q < 0$	*System Releases Heat*	*Sys --> Heat*

Some common types of Heat,

Lost: $Q = mC\Delta T \pm mL_f \pm mL_v$

Solids or Liquids $Q = mC_p\Delta T$

Gas- Constant Pressure Process $Q = mC_v\Delta T$

Gas - Constant Volume Process,

$$Q = PV \ln\frac{V_f}{V_0}$$

Gas - Constant Temperature Process $Q = 0$

Gas - Adiabatic Process

Internal Energy U

Energy Stored in a System at the Molecular Level. The System's Thermal Energy -the Kinetic Energy of the atoms due to their random motion relative to the Centre of Mass plus the binding energy (Potential Energy) that holds the atoms together.

U = Microscopic Energy contained in the System

MICROSCOPIC ENERGY

Internal energy involves energy on the microscopic scale. For an ideal monoatomic gas, this is just the translational kinetic energy of the linear motion of the "hard sphere" type atoms, and the behaviour of the system is well described by kinetic theory.

However, for polyatomic gases there is rotational and vibrational kinetic energy as well.

Then in liquids and solids there is potential energy associated with the intermolecular attractive forces.

A simplified visualization of the contributions to internal energy can be helpful in understanding phase transitions and other phenomena which involve internal energy.

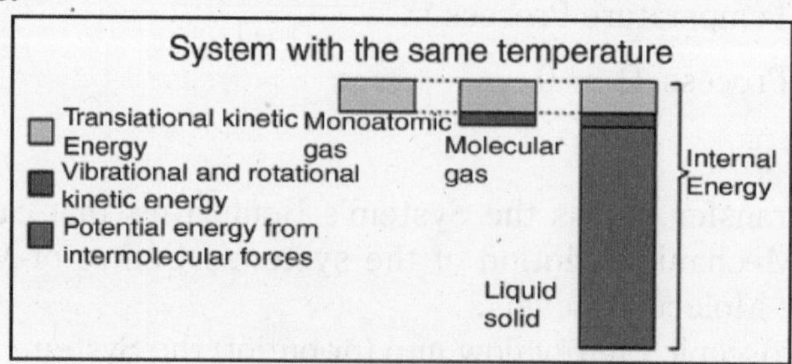

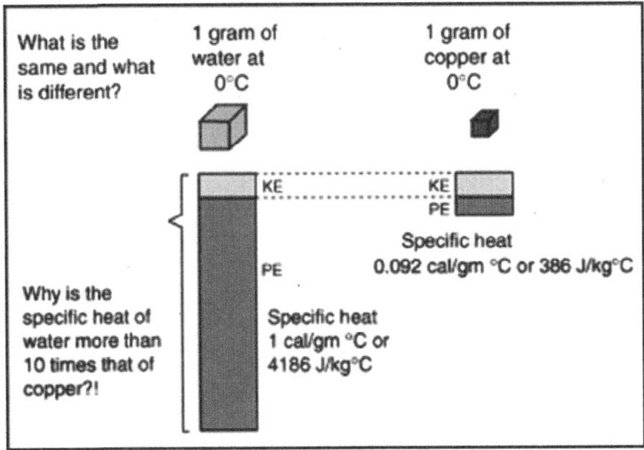

Internal Energy

When the sample of water and copper are both heated by 1°C, the addition to the kinetic energy is the same, since that is what temperature measures. But to achieve this increase for water, a much larger proportional energy must be added to the potential energy portion of the internal energy. So the total energy required to increase the temperature of the water is much larger, i.e., its specific heat is much larger. A heavy ball with an initial kinetic energy of 4000 J is trapped inside a box with rigid walls containing a cylinder constructed of small light-weight spheres. The ball crashes into the cylinder and breaks it apart. The bar graph at the right and the table at the bottom display the kinetic energy of the ball.

ENTROPY OF STEAM

The entropy diagram for steam is often convenient because it shows the relationship between

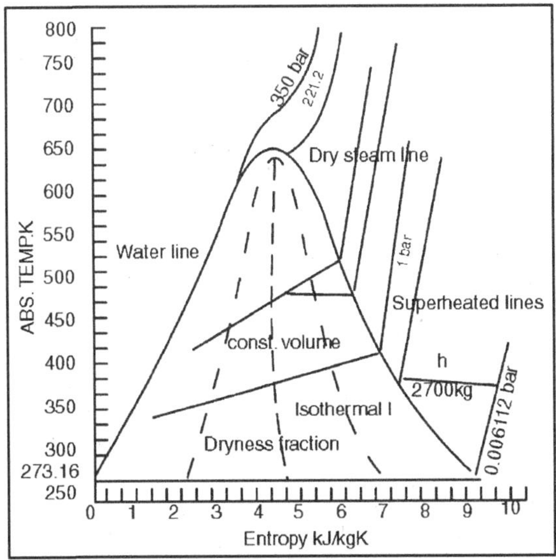

- Pressure
- Temperature
- Dryness Fraction
- Entropy

With two of the factors given - the others can be found in the diagram. The ordinates in the diagram represents the Entropy and the Absolute temperature.

The diagram consist of the following lines

- Isothermal line
- Pressure lines
- Lines of dryness fraction
- Waterline between water and steam
- Dry steam lines
- Constant volume lines

The total heat is given by the area enclosed by absolute zero base water line and horizontal and vertical line from the respective points. An adiabatic expansion is a vertical line. An adiabatic process is expansion at constant entropy with no transfer of heat.

- Critical temperature of steam is *375 to 3,380°C*
- Critical pressure is *217.8 atm*

Total Entropy of Steam

Entropy of Water

The change of entropy can be expressed as:

$dS = \log_e(T_1/T)$ (1)

where,

T = *absolute temperature (K)*

The entropy of water above freezing point can be expressed as:

$dS = \log_e(T_1/273)$ (2)

Entropy of Evaporation

Change of Entropy during evaporation:

$dS = dL/T$ (3)

where

L = *latent heat (J)*

Entropy of Wet Steam

The entropy of wet steam can be expresses as:

$dS = \log_e(T_1/273) + \zeta(L_1/T_1)$ (4)

where

ζ = *dryness fraction*

Entropy of Superheated Steam

Change of entropy during superheating can be expressed as:

$$dS = c_p \log_e(T/T_1) \qquad (5)$$

where

c_p = *specific heat capacity at constant pressure for steam (kJ/kgK)*

The entropy of superheated steam can be expressed as:

$$dS = \log_e(T_1/273) + L_1/T_1 + c_p \log_e(T_s/T_1) \qquad (6)$$

where

T_s = *absolute temperature of superheated steam*

T_1 = *absolute temperature of evaporation*

Entropy of Superheated Steam *(kJ/kgK)*

The saturated steam is exposed to a surface with a higher temperature, its temperature will increase above the evaporating temperature. The steam is then described as superheated by the temperature degrees above saturation temperature.

Absolute pressure		Saturation temperature (°C)	Steam Temperature (°C)						
(kN/m²)	*(bar)*		120	150	180	200	230	250	280
150	1.5	111.4	7.239	7.419	7.557	7.644	7.767	7.845	7.957
200	2	120.2		7.279	7.420	7.507	7.631	7.710	7.822
250	2.5	127.4		7.169	7.311	7.400	7.525	7.604	7.717
350	3.5	138.9		6.998	7.146	7.237	7.364	7.444	7.558
400	4	143.6		6.929	7.079	7.171	7.299	7.380	7.495
500	5	151.8			6.965	7.059	7.190	7.272	7.388
600	6	158.8			6.869	6.966	7.100	7.183	7.300
700	7	165.0			6.786	6.886	7.022	7.107	7.225
800	8	170.4			6.712	6.815	6.954	7.040	7.156
900	9	175.4			6.645	6.751	6.893	6.980	7.101
1000	10	179.9			6.584	6.692	6.838	6.926	7.049
1100	11	184.4				6.638	6.787	6.876	7.001
1200	12	188.0				6.587	6.739	6.831	6.956
1400	14	195.0				6.494	6.653	6.748	6.877
1600	16	201.4					6.577	6.674	6.806
2000	20	212.4					6.440	6.546	6.685
2500	25	223.9					6.292	6.407	6.558
3500	35	242.5						6.173	6.349

- Note! Steam cannot be superheated whilst it is still in the contact with water, because additional heat will evaporate more water, cooling down the superheated steam.

Superheated steam is produced by passing saturated steam through an additional heat exchanger.

Superheated steam is also called:

- Surcharged steam
- Anhydrous steam
- Steam gas

FUNDAMENTALS OF STEAM GENERATION

BOILING

The process of boiling water to make steam is a phenomenon that is familiar to all of us. After the boiling temperature is reached, instead of the water temperature increasing, the heat energy from the fuel results in a change of phase from a liquid to a gaseous state, i.e., from water to steam. A steam-generating system, called a *boiler,* provides a continuous process for this conversion.

The heat raises the water temperature, and for a specific pressure, the boiling temperature (also called *saturation temperature*) is reached, and bubbles begin to form. As heat continues to be applied, the temperature remains constant, and steam flows from the surface of the water. If the steam were to be removed continuously, the water temperature would remain the same, and all the water would be evaporated unless additional water were added. For a continuous process, water would be regulated into the vessel at the same flow rate as the steam being generated and leaving the vessel.

CIRCULATION

For most boiler or steam generator designs, water and steam flow through tubes where they absorb heat, which results from the combustion of a fuel. In order for a boiler to generate steam continuously, water must circulate through the tubes. Two methods are commonly used: (1) natural or thermal circulation and (2) forced or pumped circulation.

Natural circulation

For natural circulation no steam is present in the unheated tube segment identified as *AB*. With the input of heat, a steam-water mixture is generated in the segment *BC*. Because the steam-water mixture in segment *BC* is less dense than the water segment *AB,* gravity causes the water to flow down in segment *AB* and the steam-water mixture in *BC* to flow up into the steam drum. The rate of circulation depends on the difference in average density between the unheated water and the steam-water mixture. The total circulation rate depends on four major factors:

1. *Height of boiler.* Taller boilers result in a larger total pressure difference between the heated and unheated legs and therefore can produce larger total flow rates.
2. *Operating pressure.* Higher operating pressures provide higherdensity steam and higher-density steam-water mixtures. This reduces the total weight difference between the heated and unheated segments and tends to reduce flow rate.
3. *Heat input.* A higher heat input increases the amount of steam in the

heated segments and reduces the average density of the steam-water mixture, thus increasing total flow rate.

4. *Free-flow area.* An increase in the cross-sectional or free-flow area (i.e., larger tubes and downcomers) for the water or steam-water mixture may increase the circulation rate.

Boiler designs can vary significantly in their circulation rates. For each pound of steam produced per hour, the amount of water entering the tube can vary from 3 to 25 lb/h.

Forced circulation

For a forced circulation system, a pump is added to the flow loop, and the pressure difference created by the pump controls the water flow rate. These circulation systems generally are used where the boilers are designed to operate near or above the critical pressure of 3206 psia, where there is little density difference between water and steam. There are also designs in the subcritical pressure range where forced circulation is advantageous, and some boiler designs are based on this technology. Small-diameter tubes are used in forced circulation boilers, where pumps provide adequate head for circulation and for required velocities.

STEAM-WATER SEPARATION

The steam-water mixture is separated in the steam drum. In small, low-pressure boilers, this separation can be accomplished easily with a large drum that is approximately half full of water and having natural gravity steam-water separation. In today's high-capacity, high-pressure units, mechanical steamwater separators are needed to economically provide moisture-free steam from the steam drum. With these devices in the steam drum, the drum diameter and its cost are significantly reduced. At very high pressures, a point is reached where water no longer exhibits the customary boiling characteristics. Above this critical pressure (3206 psia), the water temperature increases continuously with the addition of heat. Steam generators are designed to operate at these critical pressures, but because of their expense, generally they are designed for large-capacity utility power plant systems. These boilers operate on the "once-through" principle, and steam drums and steam-water separation are not required.

FIRE-TUBE BOILERS

Fire-tube boilers are so named because the products of combustion pass through tubes that are surrounded by water. They may be either *internally* fired or *externally* fired. Internally fired boilers are those in which the grate and combustion chamber are enclosed within the boiler shell. Externally fired boilers are those in which the setting, including furnace and grates, is separate and distinct from the boiler shell. Fire-tube boilers are classified as vertical tubular or horizontal tubular.

The vertical fire-tube boiler consists of a cylindrical shell with an enclosed firebox. Here tubes extend from the *crown sheet* (firebox) to the upper tube sheet. Holes are drilled in each sheet to receive the tubes, which are then rolled to produce a tight fit, and the ends are beaded over. In the vertical *exposed-tube* boiler, the upper tube sheet and tube ends are above the normal water level, extending into the steam space. This type of construction reduces the moisture carry-over and slightly superheats the steam leaving the boiler. However, the upper tube ends, not being protected by water, may become overheated and leak at the point where they are expanded into the tube sheet by tube expanders during fabrication. The furnace is water-cooled and is formed by an extension of the outer and inner shells that is riveted to the lower tube sheet. The upper tube sheet is riveted directly to the shell.

When the boiler is operated, water is carried some distance below the top of the tube sheet, and the area above the water level is steam space. This original design is seldom used today. In *submerged-tube* boilers, the tubes are rolled into the upper tube sheet, which is below the water level. The outer shell extends above the top of the tube sheet. A cone-shaped section of the plate is riveted to the sheet so that the space above the tube sheet provides a smoke outlet. Space between the inner and outer sheets comprises the steam space. This design permits carrying the water level above the upper tube sheet, thus preventing overheating of the tube ends. This design is also seldom used today. Since vertical boilers are portable, they have been used to power hoisting devices and operate fire engines and tractors, as well as for stationary practice, and still do in some parts of the world.

They range in size from 6 to 75 bhp; tube sizes range from 2 to 3 in in diameter; pressures to 100 psi; diameters from 3 to 5 ft; and height from 5 to 10 ft. With the exposed-tube arrangement, 10 to 15°F of superheat may be obtained. Horizontal fire-tube boilers are of many varieties, the most common being the *horizontal-return tubular* (HRT) boiler. This boiler has a long cylindrical shell supported by the furnace sidewalls and is set on saddles equipped with rollers to permit movement of the boiler as it expands and contracts. It also may be suspended from hangers and supported by overhead beams.

Here the boiler is free to move independently of the setting. Expansion and contraction do not greatly affect the brick setting, and thus maintenance is reduced. In the original designs of this boiler, the required boiler shell length was secured by riveting several plates together. The seam running the length of the shell is called a *longitudinal joint* and is of butt-strap construction. Note that this joint is above the fire line to avoid overheating. The *circumferential joint* is a lap joint. Today's design of a return tubular boiler has its plates joined by fusion welding. This type of construction is superior to that of a riveted boiler because there are no joints to overheat. As a result, the life of the boiler is lengthened, maintenance is reduced, and at the same time higher rates of firing are permitted. Welded construction is used in modern boiler design. The products of combustion are made to pass from the grate, over the bridge wall (and under the shell), to the rear end of the boiler. Gases return through the tubes to the front end of the boiler, where they exit to the breeching or stack. The shell is bricked in slightly below the top row of tubes to prevent overheating of the longitudinal joint and to keep the hot gases from coming into contact with the portion of the boilerplate that is above the waterline. The conventional HRT boiler is set to slope from front to rear. A blowoff line is connected to the underside of the shell at the rear end of the boiler to permit drainage and removal of water impurities.

It is extended through the setting, where blowoff valves are attached. The line is protected from the heat by a brick lining or protective sleeve. A *dry pipe* is frequently installed in the top of the drum to separate the moisture from the steam before the steam passes to the steam outlet. Still another type of HRT boiler is the horizontal four-pass forceddraft packaged unit, which can be fired with natural gas or fuel oil. In heavy oil-fired models, the burner has a retractable nozzle for ease in cleaning and replacing. It is this type of design that is the most common fire-tube boiler found in today's plants. The four-pass design can be described as follows: Inside the firetube boiler the hot gases travel from the burner down through the furnace during the combustion process, and this is considered the first gas pass. The rear head of the boiler seals the flue gas in the lower portion, and the flue gas is directed to the second-pass tubes, where the flue gas flows back toward the front of the boiler.

The front head of the boiler seals the flue gas from escaping and directs the flue gas to the third-pass tubes, which causes the flow to move to the rear of the boiler. The flue gas is then directed through the fourth-pass tubes, where the flue gas moves to the boiler front and then exits to the stack. Such units are available in sizes of 15 to 800 bhp (approximately 1000 to 28,000 lb/h) with pressures of 15 to 350 psi. Some units are designed for nearly 50,000 lb/h. These units are compact, requiring a minimum of space and headroom, are automatic in operation, have a low initial cost, and do not need a tall stack. For these reasons, they find application and acceptance in many locations. Because of their compactness, however, they are not readily accessible for inspection and repairs. Larger fire-tube boilers tend to be less expensive and use simpler controls than water-tube units; however, the large shells of these fire-tube boilers limit them to pressures less than 350 psi. Fire-tube boilers serve in most industrial plants where saturated steam demand is less than 50,000 lb/h and pressure requirements are less than 350 psig. With few exceptions, nearly all fire-tube boilers made today are packaged designs that can be installed and in operation in a short period of time. Two four-pass fire-tube boilers are shown on their foundations with all associated piping and controls. Low NO*x* emissions are also critical from fire-tube boilers, as well as maintaining high boiler efficiency.

As a method for reducing NO*x* levels to as low as 20 ppm, these types of boilers can be designed using the combustion air to draw flue gas from the fourth pass. Solid fuel firing can be accommodated if there is enough space underneath the boiler to add firing equipment and the required furnace volume to handle the combustion of the fuel. The burning of solid fuels also requires environmental control equipment for particulate removal and possibly for SO2 removal depending on local site requirements. These added complexities and costs basically have eliminated fire-tube boilers for consideration when firing solid fuels.

TRIPLE POINT AND CRITICAL POINT

TRIPLE POINT

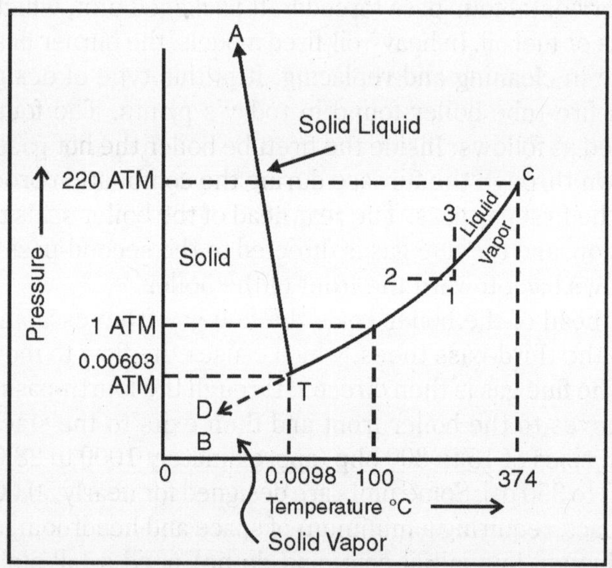

Triple point is the intersection on a phase diagram where three phases coexist in equilibrium. The most important application of triple point is water, where the three-phase equilibrium point consists of ice, liquid, and vapour. Before discussing triple point further, a basic understanding of the lines from Figure, the phase diagram of water, are first considered.

Take the line TC which gives the vapour pressure of liquid water up to the critical point C. Along this line, liquid and vapour coexist in equilibrium. At temperatures higher than that of point C, condensation does not occur at any pressure.

The line TA represents the vapour pressure of solid ice, which is a plot of the temperatures and pressures at which the solid and vapour are in equilibrium. Finally, line TB gives the melting point of ice and liquid water. The plot shows the temperatures and pressures at which ice and liquid water are in equilibrium.

Note: At the dashed line TD, liquid water can be cooled below the freezing point to give supercooled water.) The preceding paragraphs show that two phases are in equilibrium along the three solid lines. But when these lines intersect at one point C, three phases coexist in equilibrium. This intersection is the triple point, where a substance may simultaneously melt, evaporate, and sublime.

Example Problems involving the Triple Point of Water

Problem 1: Temperature vs. Pressure

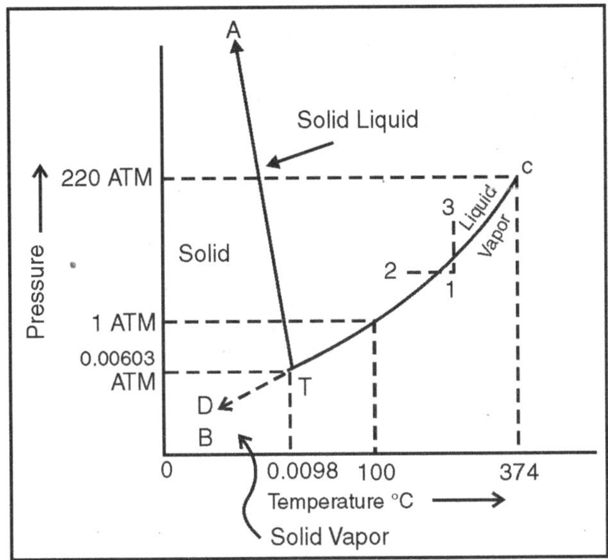

Given the phase diagram for water above, what happens to the melting point as you increase pressure?

The figure shows that as the pressure increases, the melting point increases to a maximum at the triple point. We know the temperature at this point to be zero Celsius, which is the melting point of water.

Describe the changes that occur as a result of moving across the line from point 1 to point 2. and from point 1 to point 3. In order to get to point 2, the temperature must decrease, while the pressure must increase to reach point 3. However, both process crosses the liquid-vapour equilibrium line in the direction of condensation from vapour to liquid.

Problem 2: Gibbs Phase Rule

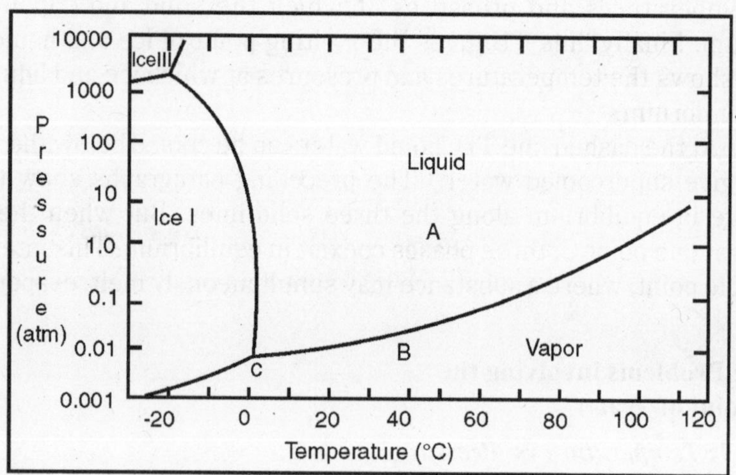

Consider the pressure-temperature phase diagram for water of Figure. Apply the Gibbs Phase Rule to specify the number of degrees of freedom at the triple point C.

This problem calls for the Gibbs Phase Rule, which is

$$P + F = C + N$$

For this system, N=2 since temperature and pressure are the only noncompositional variables. The number of components C is 1 because the system consists of solely water.

The phase rule then becomes,

$$P + F = 1 + 2$$

since we are solving for F, the equation reduces to,

$$F = 3 - P$$

Applying this point to the triple point, the number of phases present P is 3, which makes F = 0.

Now consider the points A and B, find the degrees of freedom, then compare it to point C. The phase rule F = 3 - P remains the same since it is the same system. At point A, only a single phase is present, so P=1. The number of degrees of freedom, F=2. At point B, which is the boundary between liquid and vapour phases, two phases are in equilibrium, making F = 1.

From these results we can conclude that at the triple point, where F = 0, we have no choice in the selection of externally controllable variables in order to define the system.

CRITICAL POINTS

We will discuss the occurrence of local maxima and local minima of a function. In fact, these points are crucial to many questions related to optimization problems. We will discuss these problems in later pages.

Definition

A function $f(x)$ is said to have a local maximum at c iff there exists an interval I around c such that

$$f(c) \geq f(x) \text{ for all } x \in 1$$

Analogously, $f(x)$ is said to have a local minimum at c iff there exists an interval I around c such that

$$f(c) \leq f(x) \text{ for all } x \in 1$$

A local extremum is a local maximum or a local minimum.

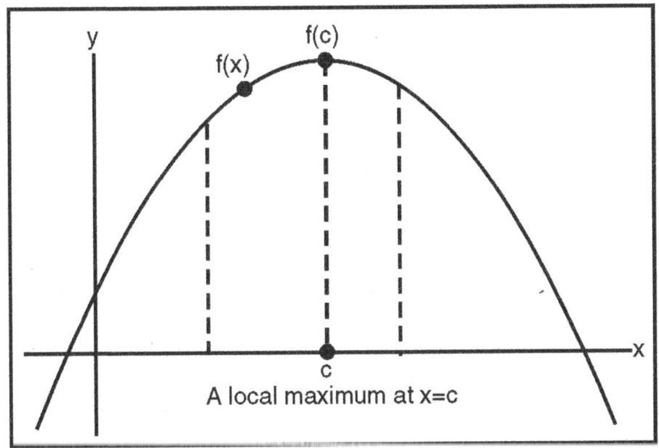

Using the definition of the derivative, we can easily show that:
If $f(x)$ has a local extremum at c, then either ,

$$f'(c) = 0 \text{ or } f'(c) \text{ does not exist}$$

These points are called critical points.

Example: Consider the function $f(x) = x^3$. Then $f'(0) = 0$ but 0 is not a local extremum. Indeed, if $x < 0$, then $f(x) < f(0)$ and if $x > 0$, then $f(x) > f(0)$.

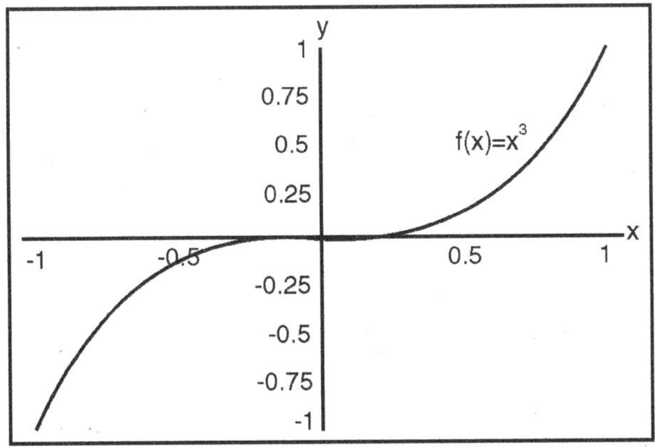

Therefore the conditions

$$f'(c) = 0 \text{ or } f'(x) \text{ does not exist}$$

do not imply in general that c is a local extremum. So a local extremum must occur at a critical point, but the converse may not be true.

Example: Let us find the critical points of,

$$f(x) = |x^2-x|$$

Answer: We have ,

$$f(x)=\begin{cases} x^2-x & \text{if } x\le 0 \\ -(x^2-x) & \text{if } 0\le x\le 1 \\ x^2-x & \text{if } 1\le x \end{cases}$$

Clearly we have,

$$f'(x)=\begin{cases} 2x\ -1 & \text{if } x<0 \\ -2x\ +1) & \text{if } 0<x<1 \\ 2x\ -1 & \text{if } 1<x \end{cases}$$

Clearly we have,

$$f'(x)=0 \text{ if } x=\frac{1}{2}$$

Also one may easily show that $f'(0)$ and $f'(1)$ do not exist.

Therefore the critical points are,

Let c be a critical point for $f(x)$. Assume that there exists an interval I around c, that is c is an interior point of I, such that $f(x)$ is increasing to the left of c and decreasing to the right, then c is a local maximum.

This implies that if $f'(x)\ge 0$ for $x\ge c$ (x close to c), and $f'(x)\le 0$ for (x close to c), then$c$ is a local maximum. Note that similarly if $f'(x)\le 0$ for $x\le c$ (x close to c), and $f'(x)\ge 0$ for $x\ge c$(x close to c), then c is a local minimum. So we have the following result:

First Derivative Test. If c is a critical point for $f(x)$, such that $f\ '(x)$ changes its sign as x crosses from the left to the right of c, then c is a local extremum.

Example: Find the local extrema of,

$$f(x) = |x^2-x|$$

Answer: Since the local extrema are critical points, then from the above discussion, the local extrema, if they exist, are among the points

$$\frac{1}{2},0,1$$

Recall that,

$$f'(x) = \begin{cases} 2x-1 & \text{if } x<0 \\ -2x+1 & \text{if } 0<x<1 \\ 2x-1 & \text{if } 1<x \end{cases}$$

(1) For $x = 1/2$, we have

$$\begin{cases} f'(x)>0 & \text{if } 0<x<1/2 \\ f'(x)<0 & \text{if } 1/2<x<1 \end{cases}$$

So the critical point $\frac{1}{2}$ is a local maximum.

(2) For $x = 0$, we have

$$\begin{cases} f'(x)<0 & \text{if } x<0 \\ f'(x)>0 & \text{if } 0<x<1/2 \end{cases}$$

So the critical point 0 is a local minimum.

(3) For $x = 1$, we have

$$\begin{cases} f'(x)<0 & \text{if } 1/2<x<1 \\ f'(x)>0 & \text{if } 1<x \end{cases}$$

So the critical point -1 is a local minimum.

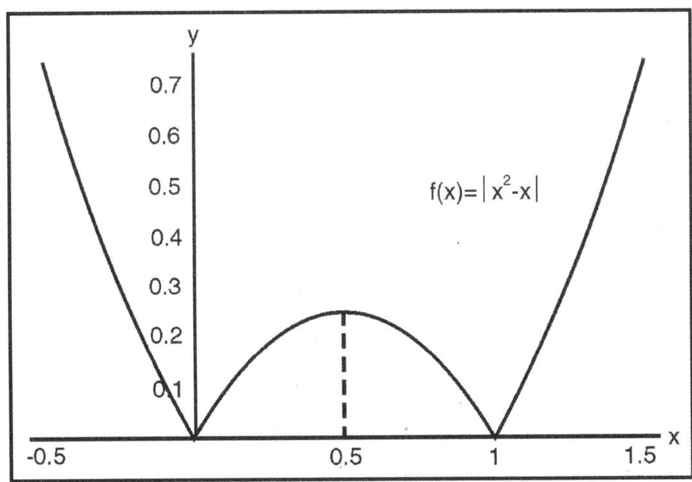

Let c be a critical point for $f(x)$ such that $f'(c) = 0$.

(i) If $f''(c) > 0$, then $f'(x)$ is increasing in an interval around c. Since $f'(c) = 0$, then $f'(x)$ must be negative to the left of c and positive to the right of c. Therefore, c is a local minimum.

(ii) If $f''(c) < 0$, then $f'(x)$ is decreasing in an interval around c. Since $f'(c) = 0$, then $f'(x)$ must be positive to the left of c and negative to the right of c. Therefore, c is a local maximum.

This test is known as the Second-Derivative Test.

Example: Find the local extrema of

$$f(x) = x^5 - 5x.$$

Answer: First let us find the critical points. Since $f(x)$ is a polynomial function, then $f(x)$ is continuous and differentiable everywhere. So the critical points are the roots of the equation $f'(x) = 0$, that is $5x^4 - 5 = 0$, or equivalently $x^4 - 1 = 0$. Since $x^4 - 1 = (x-1)(x+1)(x^2+1)$, then the critical points are 1 and -1. Since $f''(x) = 20x^3$, then

$$f''(1) = 20 > 0 \text{ and } f''(-1) = -20 < 0$$

The second-derivative test implies that $x=1$ is a local minimum and $x=-1$ is a local maximum.

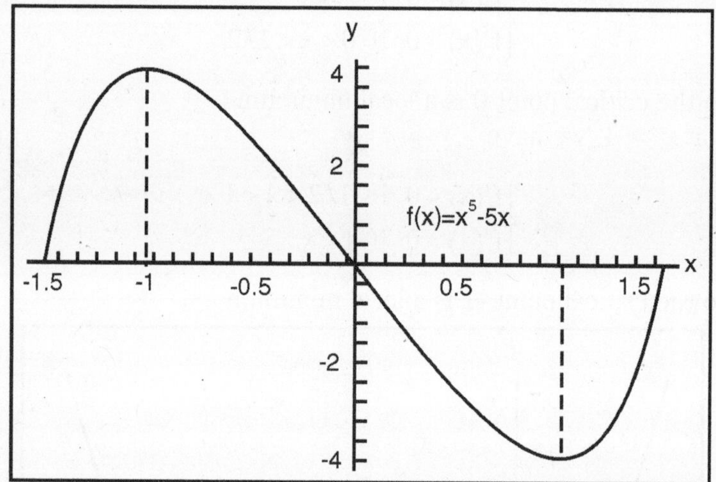

Exercise 1

Find the local extrema of,

$$f(x) = \frac{x}{1+x^2}$$

Answer to Exercise 1

First let us find the critical points. The function $f(x)$ is rational and is defined for any x. In fact, $f(x)$ is differentiable for any x.

Moreover we have,

$$f'(x) = \frac{(1+x^2) - 2x^2}{(1+x^2)^2} = \frac{1-x^2}{(1+x^2)^2}$$

So $f'(x) = 0$ implies $x = \pm 1$. Since

$$\begin{cases} f'(x) < 0 & \text{if} \quad x < -1 \\ f'(x) > 0 & \text{if} \quad -1 < x < 1 \\ f'(x) < 0 & \text{if} \quad 1 < x \end{cases}$$

Then the first-derivative test implies that $x = -1$ is a local minimum and $x = 1$ is a local maximum.

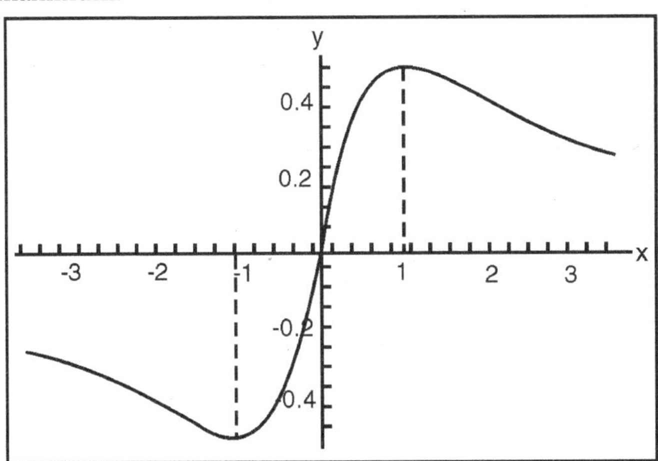

Exercise 2

Find the local extrema of,

$$f(x) = \sin(x) + \cos(x)$$

Answer to Exercise 2

Since sin(x) and cos(x) are continuous and differentiable everywhere, then $f(x)$ is continuous and differentiable everywhere.

So the critical points of $f(x)$ are the roots of ,

$$f'(x) = \cos(x) - \sin(x) = 0$$

Hence $\cos(x) = \sin(x)$ Trigonometric algebra implies that,

$x = \frac{\pi}{4} + 2n\pi$ or $x = \frac{5\pi}{4} + 2n\pi$,

where $n = 0, \pm 1, \pm 2, \ldots$ On the other hand, we have $f''(x) = -\sin(x) - \cos(x)$.

So we have,

$$f''\left(\frac{\pi}{4} + 2n\pi\right) = -2\frac{\sqrt{2}}{2} = -\sqrt{2}$$

and

$$f''\left(\frac{5\pi}{4}+2n\pi\right)=2\frac{\sqrt{2}}{2}=\sqrt{2}$$

So the second derivative test implies that,

$$x=\frac{\pi}{4}+2n\pi$$

are local maximum points and,

$$x=\frac{5\pi}{4}+2n\pi$$

are local minimum points.

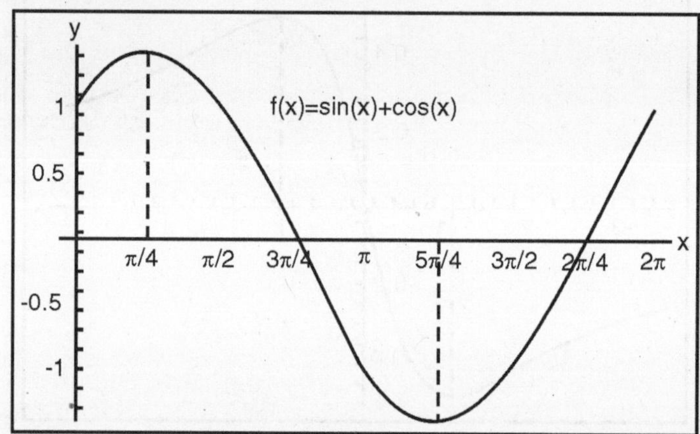

Note that the second derivative test was quite easy to use. In this case it is harder to use the first derivative test. It is sometimes hard to tell in advance which of the two tests to use. Just try one of them, and switch to the other one, if you get stuck!

CLASSIFICATION OF STEAM BOILERS

Boiler classification can be based on many factors like usage, fuel fired, fuel firing system, type of arrangement etc. Commonly known types are pulverized coal fired boilers, fluidized bed boilers, super critical boilers, oil and gas fired boilers. All cater to industrial and power generation.

DEFINITION

A boiler can be defined as a closed vessel in which water or other fluid is heated under pressure. This fluid is then circulated out of the boiler for use in various processes or power generation. In the case of power generation steam is taken out of the steam boiler at very high pressure and temperature.

In 200 B.C. a Greek named Hero designed a very simple machine which used the steam, generated in a vessel heated from below, to rotate a wheel as the steam escaped through two small pipes kept diametrically opposite, he called it as *Aelopile*.

CLASSIFICATION

Boilers vary considerably in detail and design.Most boilers may be classified and described interms of a few basic features or characteristics.Some knowledge of the methods of classificationprovides a useful basis for understanding thedesign and construction of the various types ofnaval boilers.In the following paragraphs, we haveconsidered the classification of naval boilersaccording to intended service, location of fire andwater spaces, type of circulation, arrangement ofsteam and water spaces, number of furnaces,burner location, furnace pressure, type of super-heaters, control of superheat, and operatingpressure

From 200 B.C. to date, many developments have taken place that today allow us to classify steam boilers in different ways.

Hence steam generating boilers can be classified under various categories. The main purpose of steam boilers is to generate steam, and so the way in which the steam is generated and consumed forms the major category.

The major two groups of boiler application are Industrial steam generators and power generation boilers. Boilers are also classified as fire tube and water tube boilers.

Fire tube boilers have almost become extinct; however this can be classified as:

- Locomotive boilers, which ruled rail transportation before diesel and electric engine came.
- Industrial boilers, mainly used for green projects where initial steam is required
- Domestic use boilers

Water tube boilers took over when size and capacity increased. This can be classified depending on type of circulation used to generate steam as:

- Natural circulation boiler
- Forced circulation boilers
- Super critical pressure boilers or zero circulation boilers

Depending on type of firing adopted in boilers they can be classified as:

- Stoker fired
- Pulverized coal fired
- Down shot fired
- Fluidized bed boilers
- Cyclone fired
- Chemical recovery boilers
- Incinerators

Of these the stokers which were predominantly used in early days of high pressure high capacity boilers are being replaced by pulverized coal fired boilers and fluidized bed boilers.

Stoker boilers are still designed and used in few applications like sugar industries, etc. Fluidized boilers are also going through fast development and can be now sub classified as

- Bubbling fluidized bed boilers
- Pressurized fluidized bed boilers
- Circulating fluidized bed boilers.

The higher capacity boilers are mainly circulating fluidized bed boilers due inherent limitations in bubbling bed boilers.

Boilers can be classified based on the type of fuel used as:

- Coal fired boilers
- Oil fired boilers
- Gas fired boilers
- Multi-fuel fired
- Industrial waste fired boilers
- Biomass fired boilers

Various types of arrangement are used by designers in designing the boiler for meeting the end requirement. Hence boilers are classified based on the arrangement as

- Top supported boilers
- Bottom supported
- Package boilers
- Field erected boilers
- Drum type boilers
  - Single drum
  - Bi drum
  - Three drums, but these are presently out of use
- Tower type or single pass
- Close coupled
- Two pass boilers

Boilers therefore can be classified based on firing type, fuel used, construction type, circulation type, firing system design nature, and nature of steam application.

Today's steam generating systems owe their dependability and safety to more than 125 years of experience in the design, fabrication, and operation of water tube boilers.

LANCASHIRE BOILERS

All the boilers at Pleasley were fire-tube shell types. At Teversal colliery a few years earlier, externally fired egg-ended boilers had been installed but at Pleasley it was decided to install Cornish and Lancashire boilers.

It's not certain why two different types were installed especially as the former were less efficient. It could be that the Cornish boilers were adequate for the demands of the screens and the South-pit winder which, in the early days, was not used for coal-winding, whilst the coal winding demands of the North winder would have been met by the Lancashire boilers. The Cornish ones would later be replaced by Lancashire boilers.

By 1892 there were four sets of main boilers, running at 50 psi., located on either side of the chimneys on each side of both engine-houses and feeding the winders, the screen engines and the workshop engines.

In addition, there were another four running at 50 psi, located near to the ventilation fans, feeding the fan and the dynamo engines.

The Cornish boilers were 40 ft long but the Lancashire ones were only 30 ft. A sketch plan made by the colliery manager showing the North engine-house and boilers in 1901 indicates that the boilers were arranged perpendicular to the engine-houses and that the Cornish boilers were located on the west side.

In 1900, the coal winding facilities at the South pit were upgraded and a more powerful winder was installed. The 4 Lancashire boilers in the south-east range were replaced by 5 new ones. This time they were aligned parallel with the engine-house and were flued into the side of the chimney. The coal supply was tipped from sided-door wagons on a short tightly curved railway siding. These new boilers were the recently introduced Thompson dish-ended types. They were insulated by 5 layers of bricks, but the boiler-house itself was not enclosed when the photo was taken - some time between 1900 and 1905. They were raised above the boiler-house floor with the fire-doors at about head height and firing must have been quite tricky. It is not known whether these boilers were fitted with Galloway flue tubes.

The engines supplied by these boilers were said to operate at 100 psi and they were protected by dead-weight safety-valves carrying ten plates. The pressure-gauge on the nearest boiler, however, which is venting steam from the safety-valve, although somewhat indistinct, seems to be reading at about the 10 oclock position. This seems rather low - unless the guages read up to 300 psi - which seems rather high since the boilers are unlikely to have been rated at more than 150 psi. They do not seem to have been fitted with low-water safety valves which is rather surprising. The water-level sight glasses were not protected by glass casing either - this only became mandatory in 1911.

The feed water was supplied along insulated pipes and the steam main was also well insulated although the flanges were unprotected. There was no superheating of the steam. The boilers were hand fired and the ash disposal arrangements are not obvious A new boiler-house was constructed on the east side of the South engine-house and 9 new Lancashire boilers were installed. Three existing boilers in the North engine-house east-side boiler house were retained, giving a consolidated range of twelve. Twin Unit superheaters were fitted at the rear of each flue tube giving a superheat of 90 deg F. The boilers were hand fired but it's not clear what the coal delivery and ash removal arrangements were.

Some of the dish-ended boilers installed in 1900 were transferred to Teversal and Silverhill collieries and some to the Stanton ironworks itself. Other, older boilers, were converted into exhaust-steam accumulators for the turbines, the remainder being removed altogether. A few years later, the fan-engine boilers were also removed and transferred to Silverhill colliery.

This boiler arrangement lasted until the colliery was closed. The ash removal underwent several modifications in later years and the firing arrangements were upgraded to mechanical stoking from overhead bunkers in the 1950s upgrade.

LOCOMOTIVE BOILER

It is intended that this short course will assist train crews to improve their operational skills on the foot plate by making them more aware of the capabilities of the boiler and to give a better understanding of the various parts and controls. The course will cover general boiler construction, boiler types and the various mountings attached to them with a description of their construction and correct operation. An explanation of the physical properties of steam and the physical and chemical properties of the principle fuels, coal and fuel oil. An explanation of the principles of combustion and the transformation of heat into power.

OPERATION OF LOCOMOTIVE BOILERS AND FITTINGS

The requirements of a locomotive boiler are very exacting, it having to withstand high steam pressures with a large margin of safety, combined with efficiency and economy of space. The restricted space at the disposal of the designers, and the large area of heating surface required have determined the form of the boiler. The efficiency of a boiler is measured by the amount of water that can be evaporated per lb. of coal, and this depends on the quantity of coal that can be consumed on the firegrate. For the economical production of steam, therefore, a boiler must be well designed for its work and must be handled with a degree of intelligence.

TYPES OF BOILERS AND FIREBOXES

The boiler or steam generator consists essentially of the steel shell, which includes the boiler barrel, the outer firebox wrapper plate, back plate, throat plate and smokebox tubeplate. To this is fitted the inner firebox and steel flue and smoke tubes.

Boiler for supplying superheated steam. The latter diagram illustrates a taper boiler, the cylindrical barrel is made in two sections with the larger diametre at the rear, were the barrel is joined to the outer firebox. The dome in this design, which houses the regulator valve and the auxiliary internal steam pipes, is positioned on top of the rear sloping section of the barrel, where it forms a collector for the steam above the surface of the water.

Fireboxes may be of the deep, long narrow type between the frames or of the shallow, wide type, for example, as fitted to 4-6-2 classes of locomotives. In the latter case the firebox is spread over the frames. The wide type of firebox is fitted when a large grate area is necessary.

The inner firebox is supported from the outer firebox by the foundation ring at the bottom, by crown stays at the top, and by palm stays between the firebox tubeplate and the boiler barrel. In addition, the firebox and outer wrapper plates, backplate and throatplate are stayed together with steel or copper stays, at about 4 in pitch. There are over a thousand of these stays in every locomotive boiler. Longitudinal stays are fitted between the boiler backplate and the smokebox tubeplate, and cross stays between the firebox sides above the crown. From the firebox tubeplate, the steel flue tubes, which may be anything from 1.1/2 to 2.1/4 in. diametre, pass through the boiler barrel to the smokebox tubeplate. When the boiler is fitted with a superheater, a number of large flue tubes (approx. 5 in. dia) are provided in which the superheater elements are positioned. Some boilers employ the flat top or 'Belpaire' firebox, this being designed to give a larger heating surface and a greater steam space over the firebox. The other type in general use is the round top firebox fitted to earlier designs although these were preferred by the Eastern region of British Railways until the. end of steam. It is normal practice in this country for the inner fireboxes to be made of COPPER The locomotives of the Southern region, are, however fitted with steel fireboxes and thermic syphons which improve the circulation of water around the boiler.

Copper for the inner firebox is used on account of it being a very good conductor of heat, but more especially because of its ductility and suitability for withstanding the great fluctuations of temperature within the firebox.

THE SMOKEBOX

The smokebox is an extension at the front end of the boiler barrel, which together with the blast pipe and the chimney, forms the means of producing an induced draught to the air required for combustion of the fuel in the firebox. Apart from the chimney orifice it is airtight. Other fittings in the smokebox are; The superheater header (when fitted), main steam pipes to the cylinders, Blower and ejector exhaust pipes. On some locomotives, notably Western region types the regulator valve is part of the superheater header.

The Spark Arrestor

This is fitted to the smokebox to prevent the emission of live ashes from the fire being ejected from the chimney when the engine is working hard. On some locomotives it is simply a mesh wire basket fitted between the blast nozzle and the base of the chimney known as the petticoat pipe, the larger particles being arrested and remaining in the smokebox to be removed by hand at the end of the day. A more complex type of spark arrestor has been fitted to some locomotives that are used on Railtrack lines due to the more stringent spark emission regulations. This has a vertical diaphragm plate at the back of the smokebox that breaks up the larger pieces and then deflects them towards the front of the smokebox were the flow of air is much less, the ashes are then drawn through the wire net screen before being ejected through the chimney, by which time they are small, dead and harmless.

The Superheater

The superheater is the part of the boiler designed to produce superheated steam, that is steam that is heated further after leaving the boiler and out of contact with the water from which it was generated. It is then fed directly to the cylinder valve chests. It consists of a steam collector or header for distributing steam from the boiler via the regulator valve to a series of small tubes called ELEMENTS which pass through the larger flue tubes and return to the header as superheated steam. The steam on passing through these elements has absorbed heat from the hot gases from the firebox on their way through the flue tubes to the smokebox and out of the chimney.

The superheater header is attached to the smokebox tubeplate at the outlet of the main internal steam pipe from the regulator valve and is placed horizontally across the top of the smokebox. At each side of the header are flanges for connection of the main steam pipes to the cylinders. see diagram. The Elements are fabricated from lengths of solid drawn steel tubing with three return bends, clipped together to form a bundle or element and are attached to the header with a ball and socket joint. These are known as "Melesco" elements after the company which designed them. The number of elements will vary with the degree of superheat required, however all the steam that passes through the regulator valve has to pass through the superheater elements before it can reach the cylinders.

The Brick Arch

The brick arch as its name suggests is made from high temperature refractory material, and is constructed within the firebox. It extends from the tubeplate just below the bottom row of tubes and is inclined upward for just over half the length of the firebox.

Its purpose is to lengthen the path the hot gasses have to take allowing more complete combustion of the volatile matter given off from the fuel on the firebed. When the refractory reaches its normal temperature (white hot over 2500°F) from the radiation of the firebed this further assists in the combustion of the volatile matter given off by the fuel. Secondary air passing through the firehole door is directed under the brick arch by the deflector plate and allowed to mix with the gasses on the firebed and so improving combustion. Lastly the brick arch allows the gasses to spread evenly over the tubeplate allowing the boiler to absorb as much heat as possible before being passed up the chimney.

The Firehole Door And The Deflector Plate

Various patterns of firehole door are fitted to locomotives, these give access for firing and are a method of controlling the secondary air to the firebox. Large amounts of secondary air entering the firebox is harmful to the boiler and can result in leaking tubes and stays.

Deflector Plate Or Baffle Plate

This is a metal scoop formed to the profile of the firehole and extends for about 20ins into the firebox. It is inclined downwards on the same plane as the brick arch and should point under it. A deflector plate that points over the brick arch is totally useless and must not be used. Its principle purpose is to deflect the cold incoming air under the brick arch which is white hot and preheat the air whilst allowing the air to mix with the gasses on the firebed and so assisting proper combustion. As stated above it goes a long way to preventing cold air impinging on the boiler plates causing steep temperature changes and causing leakage. Finally in the serious event of low water level and a fusible plug melting the deflector plate will prevent serious eruption of steam back through the firehole door. A correct fitting deflector plate must always be in place when the boiler is in steam.

Drop Grates, Rocking Grates And The Hopper Ash Pan

Most of our locomotives are now fitted with drop grates and hopper ashpans and are far better than the older plain firebar grates, they are there to facilitate disposal of the fire. They are of various types but the most common being the BR STD. type These consist of hinged firebars which can be controlled from the cab by means of levers. A two way stop and locking plate enables the grates to be operated with a limited amount of movement so as to break up the clinker when running, or to be rocked fully to enable the fire to be dropped completely at disposal.

A hopper ashpan is provided on most locomotives and this has hopper doors at the bottom. The doors are held shut with a catch on the side of the ashpan and are operated with the same lever as the rocking grate. The loco must not be run with the doors open and care should be taken to see that the catch is secure before moving off shed. 1 The hopper doors should always be opened before dropping the fire during disposal so as to prevent the hot fire doing serious damage to the ashpan. This operation should be done over an ashpit or other authorized cleaning point.

DAMPERS

A large proportion of the air required for combustion is admitted below the firebars and so is referred to as PRIMARY AIR. This can be regulated by the fireman by means of dampers fixed into the ashpan in the form of swinging doors. There is usually one leading and one trailing damper and are controlled by levers in the cab. The dampers should be operated in such a way as whichever direction the locomotive is travelling the opposite damper door should be used. Whenever the locomotive is about to enter a tunnel the dampers should be fully closed, the blower put on and the firehole doors closed to prevent a blowback.

PROPERTY DIAGRAMS AND STEAM TABLES

If the substance is not water vapour, the "state" of the substance is usually obtained through the use of T-s (temperature-entropy) and h-s (enthalpy-entropy) diagrams, available in most thermodynamics texts for common substances. The use of such diagrams is demonstrated by thefollowing two examples.

Example: Use of the h-s diagramMercury is used in a nuclear facility. What is the enthalpy of the mercury if its pressureis 100 psia and its quality is 70%?

Solution: From the mercury diagram, Figure A-3 of Appendix A, locate the pressure of 100 psia.Follow that line until reaching a quality of 70%. The intersection of the two lines givesan enthalpy that is equal to h = 115 Btu/lbm.

HEATING BOILER EXPANSION TANKS

Heating boiler expansion tanks are installed to absorb the initial pressure that occurs when the heating system warms up. Air molecules entrained in water inside the heatin boiler itself as well as in the heating system piping, baseboards, or radiators, expands and thus cause an initial pressure increase in the heating system.

If we do not provide some way to absorb this initial pressure increase, it is possible that the heating system's internal pressure would exceed 30 psi - the typical point at which a heating boiler pressure/temperature relief valve will open to spill excess pressure. If the relief valve is forced open in this manner the heating system will first lose water each time a heating cycle starts by heating up the boiler.

Then the heating system will take in makeup water (through the automatic water feed valve) each time the system cools down. The result would be recurrent loss and then inflow of water through the boiler, increasing the risk of system corrosion as well as wasting water and possibly causing other damage or operating problems.

When the heating boiler and system are cool, the expansion tank will contain mostly air. As the heating system warms up and air entrained in the water raises system pressure, the increased pressure forces some of the heating system water into the expansion tank, thus permitting the tank to absorb the *initial* increase in system pressure.

Naturally if the heating system is not operting properly and the boiler temperature or pressure continue to rise above acceptable levels, and since the expansion tank is a fixed volume which cannot absorb an unlimited heating system pressure increase, eventually the relief valve should also open to spill excess pressure.

TROUBLE WITH HEATING BOILER EXPANSION TANK

This means that if you see dripping at the pressure/temperature relief valve on a heating boiler, one thing to check is whether or not the expansion tank is working properly. We discuss relief valve leaks caused by expansion tank problems just below.

Expansion tanks on hot water heating systems can be divided roughly into two groups: older type bladderless heating system expansion tanks and newer type bladder-type expansion tanks such as those sold by Extrol(R).

HEATING SYSTEM EXPANSION TANKS

Older heating system expansion tanks such as the one shown in this photo need periodic service: because air in the expansion tank can become absorbed into the heating water over time, eventually the expansion tank can become waterlogged. A waterlogged heating system expansion tank will be unable to absorb the initial heating system pressure increase - it stops working. This can lead to spillage at the boiler relief valve each time the heating system warms up. Spillage at the relief valve is potentially dangerous: eventually minerals in the water can clog a leaky relief valve, causing it to stop leaking - which might look ok but this means that the relief valve has become clogged - the boiler is operating without this critical safety device and an explosion could occur.

We can determine when the expansion tank needs drainage: If the tank is heavy (try pushing it up or tapping on it) or if the relief valve is leaking, we probably need to drain the tank and let air return to it. Many expansion tanks use a special drain valve that permits air to flow into the tank as water is drained out. We drain the expansion tank simply using a garden hose connected to its drain valve. But watch out, the draining water could be hot.

Newer expansion tanks using an internal bladder should not need service. Newer type heating system expansion tanks that use an internal bladder keep their water and air separated. These tanks should not need service. If a bladder-type expansion tank has become waterlogged it's because the bladder has ruptured and the tank needs repair or replacement.

HEATING UP WITH STEAM

The amount of heat required to raise the temperature of a substance can be expressed as:

$Q = m\, c_p\, dT$ (1)

where

Q = quantity of energy or heat (kJ)

m = mass of the substance (kg)

c_p = specific heat capacity of the substance (kJ/kg ^oC) - Material Properties and Heat Capacities for several materials

dT = temperature rise of the substance (^oC)

Preferring Imperial Units - Use the Units Converter!

This equation can be used to determine a total amount of heat energy for the whole process, but it does not take into account the rate of heat transfer which is amount of heat energy per unit time

In non-flow type applications a fixed mass or a single batch of product is heated. In flow type applications the product or fluid is heated when it constantly flows over a heat transfer surface.

NON-FLOW OR BATCH HEATING

In non-flow type applications the process fluid is kept as a single batch within a tank or vessel. A steam coil or a steam jacket heats the fluid from a low to a high temperature.

The mean rate of heat transfer for such applications can be expressed as:

$q = m\, c_p\, dT/t$ (2)

where

q = mean heat transfer rate (kW (kJ/s))

m = mass of the product (kg)

c_p = specific heat capacity of the product (kJ/kg.^oC) - Material Properties and Heat Capacities for several materials

dT = Change in temperature of the fluid (^oC)

t = total time over which the heating process occurs (seconds)

FLOW OR CONTINUOUS HEATING PROCESSES

In heat exchangers the product or fluid flow is continuously heated.

The mean heat transfer can be expressed as:

$q = c_p\, dT\, m/t$ (3)

where

q = mean heat transfer rate (kW (kJ/s))

m/t = mass flow rate of the product (kg/s)

c_p = specific heat capacity of the product (kJ/kg.^oC) - Material Properties and Heat Capacities for several materials

dT = change in temperature of the fluid (^oC)

CALCULATING THE AMOUNT OF STEAM

If we know the heat transfer rate - the amount of steam can be calculated:

$m_s = q/h_e$ (4)

where

m_s = *mass of steam (kg/s)*

q = *calculated heat transfer (kW)*

h_e = *evaporation energy of the steam (kJ/kg)*

EXAMPLE - BATCH HEATING BY STEAM

A quantity of water is heated with steam of 5 bar from a temperature of *35 °C to 100 °C* over a period of *20 minutes (1200 seconds)*. The mass of water is *50 kg* and the specific heat capacity of water is *4.19 kJ/kg.°C*.

Heat transfer rate:

q = *(50 kg) (4.19 kJ/kg °C) (100 °C - 35 °C)/(1200 s)*

= *11.35 kW*

Amount of steam:

m_s = *(11.35 kW)/(2085 kJ/kg)*

= *0.0055 kg/s*

= *19.6 kg/h*

EXAMPLE - CONTINUOUSLY HEATING BY STEAM

Water flowing at a constant rate of *3 l/s* is heated from *10 °C to 60 °C* with steam at *8 bar*.

The heat flow rate can be expressed as:

q = *(4.19 kJ/kg.°C) (60 °C - 10 °C) (3 l/s) (1 kg/l)*

= *628.5 kW*

The steam flow rate can be expressed as:

m_s = *(628.5 kW)/(2030 kJ/kg)*

= *0.31 kg/s*

= *1115 kg/h*

MEASUREMENT OF DRYNESS FRACTION

To produce *100%* dry steam in an boiler, and keep the steam dry throughout the piping system, is in general not possible. Droplets of water will escape from the boiler surface. Because of turbulence and splashing when bubbles of steam break through the water surface the steam space will contain a mixture of water droplets and steam.

In addition heat loss in the pipes will condensate steam to droplets of water. Steam produced in a boiler where the heat is supplied to the water and where the steam are in contact with the water surface of the boiler - will contain approximately *5%* water by mass.

DRYNESS FRACTION OF WET STEAM

If the water content of the steam is *5%* by mass, then the steam is said to be *95%* dry and has a dryness fraction of *0.95*.

Dryness fraction can be expressed as:

$\zeta = w_s / (w_w + w_s)$ *(1)*

where

ζ = *dryness fraction*

w_w = *mass of water (kg, lb)*

w_s = *mass of steam (kg, lb)*

ENTHALPY OF WET STEAM

The actual enthalpy of evaporation of wet steam is the product of the dryness fraction - ζ - and the specific enthalpy- h_s - from the steam tables. Wet steam have lower usable heat energy than dry saturated steam.

$h_t = h_s \zeta + (1 - \zeta) h_w$ *(2)*

where

h_t = *enthalpy of wet steam (kJ/kg, Btu/lb)*

h_s = *enthalpy of steam (kJ/kg, Btu/lb)*

h_w = *enthalpy of saturated water or condensate (kJ/kg, Btu/lb)*

SPECIFIC VOLUME OF WET STEAM

The droplets of water in wet steam will occupy negligible space in the steam and the specific volume of wet steam will be reduced according the dryness fraction.

$v = v_s \zeta$ *(3)*

where

v = *specific volume of wet steam* (m^3/kg, ft^3/lb)

v_s = *specific volume of the dry steam* (m^3/kg, ft^3/lb)

ENTHALPY AND SPECIFIC VOLUME OF WET STEAM

Steam at pressure *5 bar gauge* has a dryness fraction of *0.95.*

Total enthalpy can be expressed as:

h_t = *(2085 kJ/kg) 0.95 + (1 - 0.95) (670.4 kJ/kg)*

= *2,014 kJ/kg*

Specific volume can be expressed as:

$v = (0.315\ m^3/kg)\ 0.95$

$= 0.299\ m^3/kg$

DRYNESS FRACTION VARIATION

The density of a cubic metre of wet steam is higher than that of a cubic metre of dry steam. If the quality of steam is not taken into account as the steam passes through the flowmetre, then the indicated flowrate will be lower than the actual value.

To reiterate; dryness fraction is an expression of the proportions of saturated steam and saturated water. For example, a kilogram of steam with a dryness fraction of 0.95, contains 0.95 kilogram of steam and 0.05 kilogram of water.

Example: As a basis for the following examples, determine the density of dry saturated steam at 10 bar g with dryness fractions of 1.0 and 0.95.

Dryness Fraction(x) = 1.0
Specific Volume of dry steam (V_g)
at 10 bar g (from steam tables) = 0.1773 m³/Kg

$$\text{Density } (\rho) = \frac{1}{0.1773\,\text{m}^3/\text{Kg}}$$

with x having a dryness fraction
of 1.0, density (ρ) = 5.6414Kg/m³
Dryness fraction (x) = 0.95
Specific volume of dry steam (V_g)
at 10 bar g (from steam tables) = 0.1773 m³/Kg
Spcific Volume of Water (V_f)
at 10 bar g (from steam tables) = 0.0011

- Volume occupied by steam @x = 0.95 = 0.95×0.1773 = 0.1684 m³
- Volume occupied by Water @x = 0.95 = 0.05×0.0011 = 0.000055m³
- Total volume occupied by steam and water = 0.1684 + 0.000055m³ = 0.168455m³

Density (ρ) of mixture,

$$= \frac{1}{0.168455\,\text{m}^3/\text{Kg}} = 5.9363\,\text{Kg/m}^3$$

Difference in density = 5.9363 Kg/m³ – 5.6414 Kg/m³ = 0.2949 Kg/m³

Therefore, a reduction in volume is calculated to be 4.97%,

$$\text{Density of steam} = \frac{1}{v_g x}$$

$$\frac{\text{Indicated massflowrate}}{\text{Actual flowrate}} = \frac{\sqrt{\text{Density at calibrated dryness fraction}}}{\sqrt{\text{Density at actual dryness fraction}}}$$

Equation

The proportion of the volume occupied by the water is approximately 0.03% of that occupied by the steam. For most practical purposes the volume occupied by the water can be ignored and the density of wet steam can be defined as shown in Equation

Where:

Vg = Specific volume of dry steam

x = Dryness fraction

Using Equation 4.4.3, find the density of wet steam at 10 bar g with a dryness fraction (x) of 0.95. The specific volume of dry steam at 10 bar g (g) = 0.1773 m3/kg. This compares to 5.9363 kg/m3 when calculated as a mixture.

Effect of Dryness Fraction on Flowmetres

To reiterate earlier comments regarding differential pressure flowmetre errors, mass flowrate (qm) will be proportional to the square root of the density, and density is related to the dryness fraction. Changes in dryness fraction will have an effect on the flow indicated by the flowmetre.

Equation can be used to determine the relationship between actual flow and indicated flow:

$$\frac{\text{Indicated flowrate}}{\text{Actual flowrate}} = \frac{\sqrt{\text{Density at } x = 1.00}}{}$$

$$\frac{1 \text{ Kg/s}}{\text{Actual flowrate}} = \frac{\sqrt{5.6414}}{5.9294} = \frac{2.3752}{2.4350}$$

Actual flowrate = 1.0252 Kg/s

$$\text{Per centage error} = \frac{\text{Indicated flow} - \text{Actual flow}}{\text{Actual flow}} \times 100\%$$

$$\text{Per centage error} = \frac{1 - 1.0252}{1.0252} \times 100\% = -2.46\%$$

Equation: All steam flowmetres will be calibrated to read at a pre-determined dryness fraction, the typically value is 1. Some steam flowmetres can be recalibrated to suit actual conditions.

Example: Using the data from Example determine the per centage error if the actual dryness fraction is 0.95 rather than the calibrated value of 1.0, and the steam flowmetre was indicating a flowrate of 1 kg/s. Therefore, the negative sign indicates that the flowmetre under-reads by 2.46%.

Equation is used to compile the graph shown in Figure:

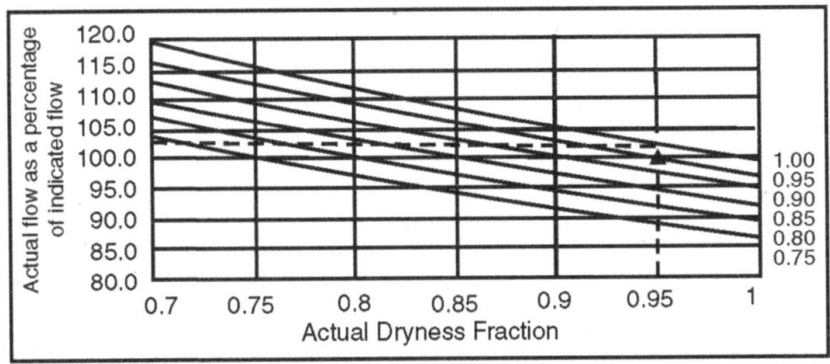

Fig. Effect of Dryness Fraction on Differential Pressure Flowmetres

Effect of Dryness Fraction on Vortex Flowmetres

It can be argued that dryness fraction, within sensible limitations, is of no importance because:

- Vortex flowmetres measure velocity.
- The volume of water in steam with a dryness fraction of, for example, 0.95, in proportion to the steam is very small.
- It is the condensation of dry steam that needs to be measured.

However, independent research has shown that the water droplets impacting the bluff body will cause errors and as vortex flowmetres tend to be used at higher velocities, erosion by the water droplets is also to be expected. Unfortunately, it is not possible to quantify these errors.

BOILER MOUNTINGS AND BOILER CONTROLS

In addition to the boiler being of sound design and construction to be able, to withstand the forces of steam pressure within, certain safety features or safety devices must be. fitted to it to prevent the operating conditions becoming dangerous through over pressure or low water level. Both of these conditions could present a very serious hazard to anyone within a very wide area and in the event of a boiler explosion would cause catastrophic damage to the surrounding area.

On the next page, study the photographs taken in the aftermath of a boiler rupture. These pictures show what incredible forces are stored up within a locomotive boiler, and if allowed to get out of control and the boiler plates fail, anyone on the footplate would be killed instantly or at the very least, very badly scalded.

Fortunately these incidents are rare nowadays due to more controlled inspections, however the locomotive boiler is still under full manual control and should be under constant supervision when in steam.

By way of another example to illustrate the enormous power stored up within the locomotive boiler, consider the following simple calculation.

The average locomotive boiler backplate is approx 60in wide and 84in high and the boiler pressure is, say, 2501b in.

$60 \times 84 = 5040$ sq. inches

$5040 \times 250 = 1{,}260{,}000$ lbs.

$1{,}260{,}000\ 2{,}240 = 562.5$ TONS

This is the steam pressure in front of you when you are stood on the footplate. Just something to consider when you are next booked out on a loco.

Safety Valves

The safety valves are perhaps the single most important safety device fitted to the boiler. They are a mandatory requirement and are fitted to prevent the boiler pressure from exceeding the registered working pressure of the boiler.

This is the steam pressure for which it was designed and is indicated by a metal tablet secured to the firebox backplate.

The "Ross Pop" type of safety valve was in extensive use on the former British Railways and it is still in use today. In this design, when the working pressure is reached, the spring loaded valve rises and admits a small amount of steam through the seat of the valve into the Annular Chamber, this gives increased lift to the valve against the pressure of the spring The steam then escapes into the body of the valve and escapes through the holes on the top cap. The steam, when escaping acts on the increased area of the top cap, forcing the valve open still further until such time as the pressure in the boiler has dropped slightly. The spring then overcomes the pressure of the escaping steam and closes the valve instantaneously with a "pop" action.

Blowing off is wasteful of steam and can be avoided by careful management of the fire and injectors. For example: On a large 4-6-0 locomotive, for each minute the safety valve is blowing there is a loss of water of approximately 15 gallons and a loss of fuel of approximately 10 lb.

Water Gauges

Probably the most serious condition any steam boiler can suffer, is low water level. The crown of the firebox must always be covered with water at all times when the boiler is in use. If it should become uncovered, even for a short time, the firebox roof will very quickly become red hot. When this happens the metal it is made of becomes soft and plastic, the steam pressure then forces it off the supporting stays disgorging the remaining steam and water contents into the firebox with violent force. As it is not possible to see through the steel plates, some form of visible water indication is required, and this is performed by the water gauge. The designs consist typically of a top steam cock "K', a bottom water cock 'V' and a drain or blow through cock "C". Cocks "X' and "B" are connected together with a glass tube called the gauge glass. Fitted around the gauge glass is a protector, which is made from thick toughened glass, this protects the glass from being damaged and also protects the train crew in the event of a burst gauge glass. The water level in the boiler is shown inside the glass tube. With the boiler in steam. The water level must always be visible. With it showing in the bottom nut the firebox will still be covered. The Perforated plate behind the protector usually has on it a set of diagonal black fines, with NO water showing in the glass these black lines will appear the same. With water showing, the part that contains water will be refracted and the black lines will appear to show the other way round. This is one way to tell whether there is water in the glass or not (i.e. Completely full or empty).

Testing Gauge Glass

The gauge glass must be tested at least once after taking over the locomotive and at regular intervals during the time that you have control of the boiler The important thing to remember when testing the gauge glass is that from the top cock "X' the steam should HISS. S. And from the bottom cock "B" the water should ROAR. When it comes out of the drain pipe.

How to change a leaking or burst gauge glass:

- If the gauge glass bursts, throw a coat or sack over the effected fitting, and shut off the steam cock "X' and water cock 'M" instantly
- Open the drain cock "C"
- Remove the perforated backplate and remove the protector
- Remove the top plug from the top fitting and remove the loose restrictor
- Remove the gland nuts from the top and bottom fitting and clean out any remaining glass and rubber washers
- Place the new glass in from the top (if possible) feeding the parts on to the glass as follows:
  - Top rubber washer
  - Top gland follower
  - Top gland nut
  - Bottom gland nut
  - Bottom gland follower
  - Bottom rubber washer
- Push the glass firmly down into the bottom fitting
- Feed in the rubbers, gland followers and finally screw up the gland nuts HAND TIGHT only
- Replace the restrictor and the top plug in the top fitting
- Refit the protector and the perforated back plate, and ensure it is secure
- Warm the glass through by opening the steam cock "X' slowly
- Slowly close the drain cock "C" (the glass may burst again)
- Finally open the water cock 'S" and check the water level rises in the glass.

Boiler Pressure Gauge

In order to detect the steam pressure within the boiler, a pressure gauge is fitted. This is a mandatory requirement and is always fitted directly to the boiler. A cock is usually fitted to allow changing of the gauge should it become defective. The design is always of the Bourdon tube type, it made from phosphor bronze.

The dial has a scale and is usually marked off in Pounds per Square Inch. (LBf/in2) though sometimes it may have a dual scale marked in Bars, usually in increments of ten. On every boiler gauge there is marked a Red fine on the scale, this is the maximum working pressure of the boiler and the point at which the safety valves will lift.

Fusible Plugs

In order to give the fireman a warning and to some degree protect the crown of the firebox, One or more fusible plugs are screwed into the roof of the firebox. These are made from bronze and have a Lead core which will melt at quite a low temperature. If the water level drops too low and uncovers the plugs, the lead melts and allows steam to escape into the firebox. The hole the lead fills is only about 318 in. Dia. And so will not be sufficient to extinguish the fire.

Blower Valve and Ring

The blower consists of a perforated ring or pipe fitted round the top of the blast pipe and is connected to a steam supply from the boiler via a valve in the cab and is usually under the control of the driver. The function of the blower is to create a vacuum in the smokebox for the following;

- To increase the draught on the fire when the locomotive is stationary, or in order to raise steam pressure.
- To clear smoke.
- To counteract back draught in the case of 3. Whilst working a train or light engine, the blower valve must always be opened prior to closing the regulator, or before entering a tunnel.

Blowdown Valvc

In order to remove the contaminants of the boiler water which has allowed to settle as sludge at the bottom of the foundation ring, a Blowdown valve is fitted. When this is opened the boiler pressure forces out this sludge with violent but controlled force. Before operating the Blowdown valve the boiler water level must be high in the glass and have sufficient steam pressure to operate the injectors to be able to restore the water level when the valve is closed.

The water level must not be allowed to drop out of sight in the gauge glass, and must be continuously watched when blowing down, with the boiler in steam. The operating handle should be secured when not in use to prevent tampering or inadvertent operation. Care must also be taken when sighting the locomotive for blowing down and a clear warning given to people nearby who may be scalded from the outlet pipe.

REGULATOR VALVE

The regulator valve is probably the biggest boiler fitting and is positioned usually within the dome of the boiler. It is operated by the handle in the cab, mounted on the backplate where it passes through a gland into the boiler. There are several types of valve all designed to provide a controlled passage of steam from the boiler through the superheater (if fitted) to the cylinders. The most common type is the Vertical Slide which has two, three or four ports on the valve (sometimes called Second valve), and two ports on the pilot valve (First valve). The pilot valve opens first when the regulator handle is moved which gives controlled starting. The other types of regulator valve are: Horizontal slide type (fitted to many Ex LMS classes); the Double beat type (fitted to many Ex LNER classes); and the Balanced piston type which are fitted to all the EX SR West Country, Battle of Britain and Merchant Navy classes. Some types of regulator are fitted in the smokebox within the superheater header.

RANKINE CYCLE

DEFINITION

The Rankine cycle is a thermodynamic cycle used to generate electricity in many power stations, and is the real-world approach to the Carnot cycle. Superheated steam is generated in a boiler, and then expanded in a steam turbine. The steam turbine drives a generator, to convert the work into electricity. The remaining steam is then condensed and recycled as feed-water to the boiler. A disadvantage of using the water-steam mixture is that superheated steam has to be used, otherwise the moisture content after expansion might be too high, which would erode the turbine blades.

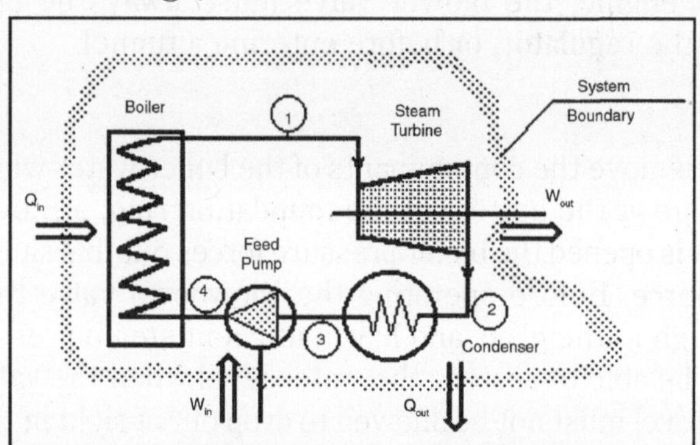

Rankine cycle is a heat engine with vapour power cycle. The common working fluid is water.

The cycle consists of four processes:

- 1 to 2: Isentropic expansion (Steam turbine)
- 2 to 3: Isobaric heat rejection (Condenser)
- 3 to 4: Isentropic compression (Pump)
- 4 to 1: Isobaric heat supply (Boiler)

Work output of the cycle (Steam turbine), W1 and work input to the cycle (Pump), W2 are:

$$W1 = m\,(h1-h2)$$
$$W2 = m\,(h4-h3)$$

where m is the mass flow of the cycle. Heat supplied to the cycle (boiler), Q1 and heat rejected from the cycle (condenser), Q2 are:

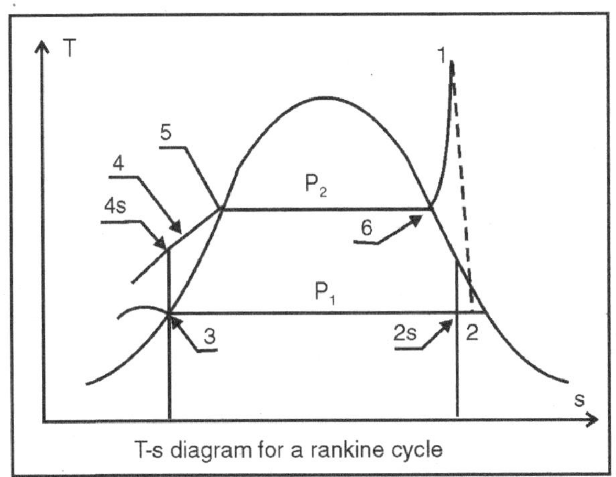

T-s diagram for a rankine cycle

Q1 = m (h1-h4)

Q2 = m (h2-h3)

The net work output of the cycle is:

W = W1 - W2

The thermal efficiency of a Rankine cycle is:

$$\eta = W/Q_1$$

The efficiency of the Rankine cycle is not as high as Carnot cycle but the cycle has less practical difficulties and more economic

THE BASIC RANKINE CYCLE

We will compare our regenerative cycle to a typical Rankine cycle. The Rankine cycle to which we will compare has the following parametres:

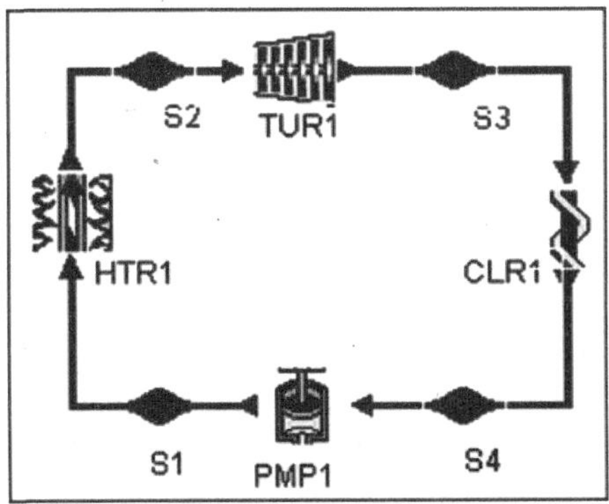

Fig. A Rankine cycle

The operating limits are:

- Heater pressure of 5 MPa
- Heater exit temperature of 400 C
- Cooler pressure of 10 kPa

and its efficiencies are:·

- Carnot efficiency: 52.6%
- Thermal efficiency: 36.2%

The regeneration cycle we examine will operate under the same limits. For reference, the Rankine cycle layout is shown below.

RANKINE CYCLE WITH RENGERATION

Improving Cycle Efficiencies

Improving cycle efficiency almost always involves making a cycle more like a Carnot cycle operating between the same high and low temperature limits.

The Carnot cycle is maximally efficient, in part, because it receives all of its heat addition at the same temperature, which is the highest temperature in the cycle. Similarly, it rejects all of its heat at the same low temperature. The T-s diagram below details the working of a Carnot cycle operating between the same temperature limits as our Rankine cycle.

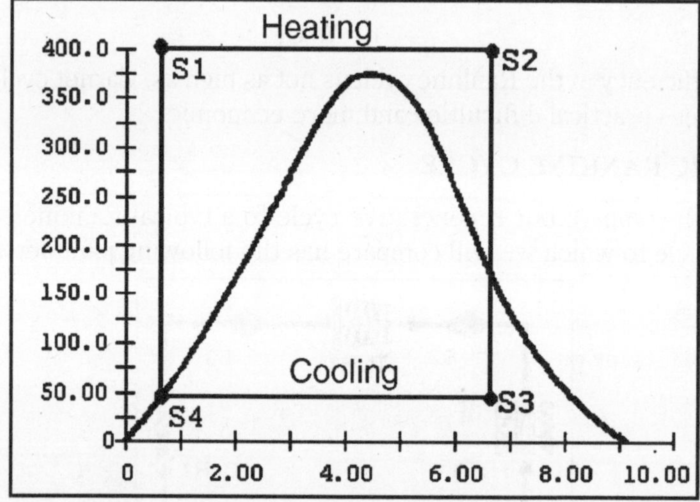

Fig. Carnot Cycle T-s Diagram

Most cycles don't have all of their heat addition or rejection at one temperature. So, when we look to improve a cycle's efficiency, we often consider the *mean temperature of heat addition,* T_a and *the mean temperature of heat rejection,* T_r. These reflect what the temperature would have been if the same amount of heat had been added (or rejected) all at one temperature.

They allow us to treat improving cycle efficiencies as we would for a Carnot cycle: by raising T_a or lowering T_r. For reversable heat transfer, the average temperature of heat addition is,

$$T_a = Q_{in} / \Delta S$$

and the average temperature of heat rejection is,

$$T_r = Q_{out} / \Delta S$$

For more efficient cycles, we would like to add heat at a higher temperature and reject it at a lower temperature.

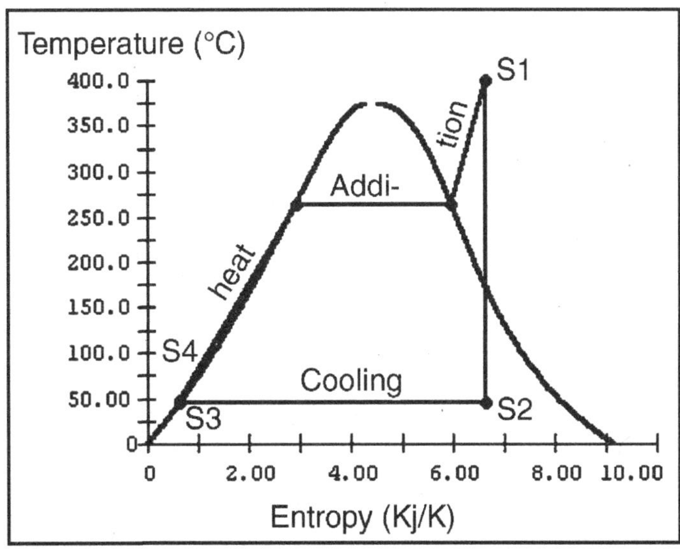

Fig. Rankine Cycle T-s Diagram

Knowing this, let's look at Figure. In the Rankine cycle, the above equations tell us we are adding heat between states S4 and S1 at an average temperature of about 226.7 C. The heat rejection from S2 to S3 occurs at the cooler saturation temperature of 45.8 C. As a quick check, we can find the Rankine cycle's thermal efficiency by applying the relation for Carnot efficiency to the mean Rankine cycle temperatures:

$$\eta = (T_a - T_r) / T_a = (226.7 - 45.8) / (226.7 + 273.15) = 36.2\%$$

Which is the same answer we get applying the usual $\eta = W_{net} / Q_{hi}$ relation.

How regeneration works

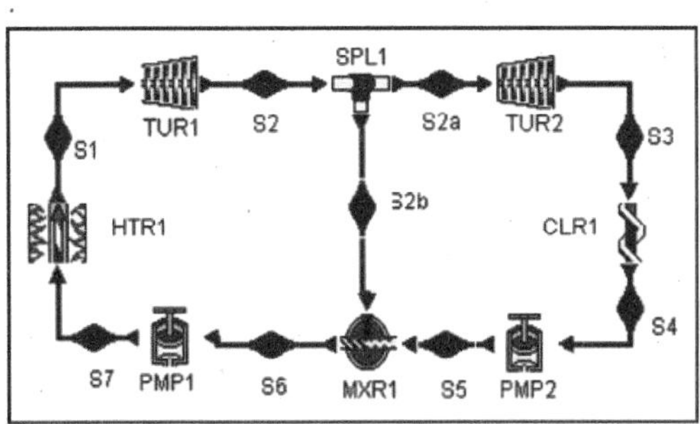

Fig. A Rankine Cycle with Regeneration

The idea behind regeneration is that we split the turbine into high-pressure and low-pressure stages and do the same for the pump. Then, we can divert some of the heat in the fluid as it leaves the high-pressure turbine and add it to the cool fluid leaving the low-pressure pump, thereby sending fluid with a higher temperature to the heater.

CHOOSING REGENERATION ASSUMPTIONS

With the design layout complete, we turn to adding the assumptions which allow *CyclePad* to solve the cycle. For this example, there are four places where we have to make new assumptions: the outlet of the high-pressure turbine (S2), the splitter (SPL1), the outlet of the low-pressure pump (PMP2), and the inlet of the high-pressure pump (PMP1). We will look at each of these in turn.

The High-Pressure Turbine Outlet (S2)

What pressure do we choose for the extracted feedwater? We don't know yet, so we'll choose 200 kPa, which gives the two turbines pressure ratios of 25 and 20. This makes them roughly equal and keeps either one from having an astronomically high pressure ratio. (The original Rankine cycle had a turbine PR equal to 500!) Later, when we have the cycle solved and we can let *CyclePad* do sensitivity analyses, we will see if another pressure works better.

The Splitter (SPL1)

The splitter is used to draw some of the working fluid from the high-pressure turbine stage and direct it towards the mixer. The assumption we make here is that the splitter is isoparametric. This means that the stuff exiting the splitter is the same as the stuff entering it.

Quick Note

The other assumption is that the splitter is not isoparametric. This simulates situations when we use a special splitter that allows us to separate the saturated mixture into two streams that each have different proportions of liquid and vapour. For instance, we could split 1 kg/sec, 60% quality stream into a pair of 0.4 kg/sec, 0% quality and a 0.6 kg/sec, 100% quality streams. This allows each stream to have different specific properties (v, h, and so on), though they still have the same temperature and pressure.

We might think that this would be advantageous to regeneration so that we could send only saturated vapour (which has higher enthalpy) to the low-pressure turbine and get more work out of it and send the low quality remainder down to heat the water entering the high-pressure pump. So why not do this? Let's try it. We can retract the isoparametric assumption and instead assume the stuff entering the low-pressure turbine is a saturated vapour. Then when our other assumptions for this cycle have been made, we will see that the efficiency has changed by a few hundredths of a per cent and we see that such an approach does not improve cycle efficiency by much.

Our high-pressure pump condition (S6 below) requires that the total enthalpy H (not h) entering the high-pressure pump is constant, so the lower quality of the steam entering the mixer requires a higher mass flow into the mixer, nearly negating the benefit of the higher h steam that enters the low-pressure turbine.

That is, we could send lower h stuff to the pump, but we'd have to send more of it. Of course, any efficiency improvement is better than none, but it isn't always realistic to assume that we can use the special splitter than separates vapour from liquid, so we'll continue using the isoparametric splitter.

The Low-Pressure Pump Outlet (S5)

What pressure should the water at this state have? This water, which enters the pump at the cooler pressure, needs to be pumped up to the pressure of the water extracted from the high-pressure turbine. This is a matter of simple hydrostatics: if we make it lower, the higher pressure extract water will flow backward through the low-pressure pump and, if we make it higher, the low pressure pump water will flow backward through the splitter.

We could just set this pressure at 200 kPa, the pressure we set at the high-pressure turbine outlet. However, the better approach is to tell *CyclePad* to make the two pressures *equal*, using the "Equate T(S5) to another parametre" option. This way, when we need to experiment with different feedwater pressures, we don't need to make the change in two places. Another equivalent solution is to declare MXR1 to be isobaric.

The High-Pressure Pump Inlet (S6)

Our whole purpose in adding heat to this water is to raise its temperature before it enters the heater and improve the cycle efficiency. The water exiting the low-pressure pump is only at 46 C and the water entering the heater in the original Rankine cycle was at about the same temperature (adding pressure to an incompressible fluid doesn't raise its temperature much). How high can we heat the water to improve this?

The only limit we really need to consider is the practical use of the pump. We make sure the water entering the pump in a simple Rankine cycle is a saturated fluid because pumps cannot handle vapour very well. We have the same consideration here. We want to heat this water up as much as we can, but not so much that some of it starts to vaporize again. We recall that this is the state of a saturated liquid. This is our assumption for the high-pressure pump inlet: it is saturated with quality $=$ 0.

EXAMINING REGENERATION EFFICIENCIES

EFFECT ON CYCLE EFFICIENCY

We now have made the changes needed to add a regeneration stage to our Rankine cycle. Here are the new efficiencies and the ones of the plain Rankine cycle for comparison. Also shown is the per cent gain in power output of the modified cycle when supplied with the same heat as the unmodified Rankine cycle and the mean temperatures of heat addition for both cycles.

Cycle Efficiencies	Unmodified Rankine Cycle	Rankine Cycle with Regeneration
Carnot efficiency	52.6%	52.6%
Thermal efficiency	36.2%	38.4%
% Increase in Power Generation	0%	6.1%
mean temperature of heat addition	226.7 C	251.5 C

All bodies in nature are presented to us in one or other of the three differernt forms or states- viz. the solid, the liquid and the gaseous. Most bodies can be made, by changes in temperature, to assume thses different forms. The simplest and best know example is that of water. At the ordinary temperature of out climate, this is a perfect liquid; at a considerably lower temperature, it becomes solid, and then it is called ice; at a very high temperature, it becomes gaseous, and is then called steam.

A liquid, under the influence of heat, generates an *aeriform* fluid; this fluid, when free to expand in a given space, is called a vapour, but when it completely fills that space, and prevents the futher vapourization of that fluid, it is called a gas or an elastic fluid.

Although simple variations of temperature and volume will cause a gas to pass into the state of vapour, and vice versa, yet the physical properties of the fluid, in these two states, are essentially different.

Thus, by heating a given volume of air, or any other vapour, from 32 degrees to 212 degrees Fahrenheit, it's elastic force is increased only in the ratio of 1 to 11, or of 8 to 11, nearly; whereas, the elastic force of the vapour of water, while in contact with the generating liquid, is increased in the ratio of about 1:150.

It is this enormous development of force, produced by the application of heat, which accomplishes in modern times, the amazing mechanical effects of steam, when employed as a moving power.

The Ancients had no such clear notions of the nature of steam as those which have just been stated; although, in the writings of Plato, we find speculations concerning the nature and properties of the four elements, as they were called, Earth, Air, Fire and Water, which show that they were partially acquainted with the subject. It is true that only about a century and a half have elapsed, since steam began to be usefully employed as a moving power; and yet, the knowledge of it's expansive force can be traced to remote antiquity. This discovery must, indeed have forced itself on the notice of those who first made use of a pot, with a lid on it, in culinary operations.

It is, therefore, simply ridiculous to refer the discovery of the properties of *steam*, or the invention of the steam engine, to the Marquis of Worcester, who flourished in 1663, because he observed the lid of his tea-kettle thrown off by the force of the steam.

For aught we know to the contrary, Adam and Eve may have witnessed such a phenomenon as this, in their post-paradisical state.

Hero of Alexandria, who flourished around 120 B.C., in a Treatise on Pneumatics, written in Greek, records the principal facts that were then known regarding the vapour of water, and gives an account of several machines in which the force of steam is employed to produce motion.

One of these, the 45th in the book, consists of a pot with a lid, into which is fastened an upright tube, terminating in a perforated hemispherical cup, in which is placed a movable ball. Water having been put into the pot, and the whole apparatus placed over a fire, as soon as the vapour or steam issues with sufficient force through the tube, it causes the ball to dance over the cup, with a rapidity varying according to the force of the steam.

It is a representation of this apparatus, which is called in Greek *lebes*, that is, pot or kettle; it is taken from Commandine's edition of the work of Hero, published at Urbino in 1575, and has evidently been copied from the oldest manuscript.

The records of the Steam Engine Maker's Society are a unique source, providing detailed information about many aspects of individual lives from 1835 to the First World War; details concerning unemployment, sickness, aging and migration which are available from no other source. However, they concern not some 'random sample' from the country's population but the membership of a particular trade union. If we wish to use the SEM data to draw conclusions which relate to some wider group, we must understand the particular place of SEM members within the union movement, the engineering industry, and the wider society they lived in.

This document contains:

- A brief outline of the union's history
- The known characteristics of its membership
- A description of the benefit system from which most of the records derive:
  - Travel Pay
  - Sick Pay
  - Superannuation
  - Unemployment Benefit
  - Other Benefits

Unemployment Pay

Unemployment benefit on modern lines, meaning a weekly payment rather than the aid to job-search provided by travelling benefit, was introduced by the SEM's London branch in 1838 without consulting the rest of the union. This was formalised the following year, London members paying increased contributions to cover the cost. The first outside London listing unemployment benefit payments was Manchester, in the 1844-5 *Annual Report*, and by 1848-9 it was being paid by the branches in London, Manchester, Woolwich, Greenwich, Romford, Stratford, Southampton, Patricroft, and Bristol. The SEM introduced unemployment pay comprehensively following the 1848 recession, when it was seeking to re-establish itself following the defection of large numbers of members to the newly-formed ASE; the ASE took most of its leadership and methods of operation from the *Journeymen* Steam Engine Makers' Society (JSEM), which had always relied on unemployment pay rather than tramping.

Other Benefits

Other benefits were an accident grant of fifty pounds for any member 'rendered incapable of following his employment, either by loss of limb or eyesight' a separate 'Contingent Fund' created in the 1870s to provide strike pay, and a discretionary Benevolent Fund which would seem to have been the only provision for widows and orphans. In return for these benefits, members paid two shillings and three pence per month subscription in 1846, raised to three shillings per month in 1851; this was significantly less than the ASE, which charged a shilling per week. There was also an admission charge dependent on age, varying from 10 shillings and six pence for men under 23 to five pounds for men aged 45; men aged over 45 could not join.

FIRE TUBE BOILER

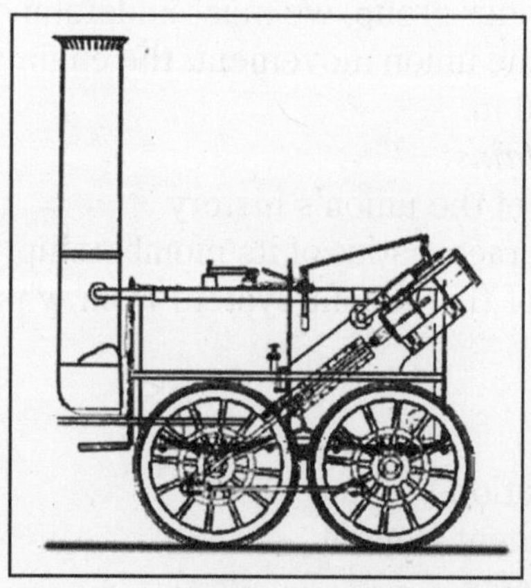

Fire-tube boilers are those device in which hot gases, which originates from the fire passes through one or more tubes within the boiler. After three main historical forms of boilers (low-pressure tank or haystack boilers, tube boilers, high-pressure water-tube boilers), fire tube boilers are manufactured. In these boilers, hot gases pass through one or more tubes running through a sealed container of water. According to thermal conduction, the heat energy of the gases pass away from the sides of the tubes, resulting into the heating of water and finally steam.They are important boilers that are available in both horizontal and vertical configurations. They are also some times referred as smoke-tube boiler.

Construction and Usage

Usually, the tank of the boiler is cylindrical in shape. It can also be either vertical or horizontal. Also regarded as shell boiler, smoke tube boiler and fire pipe, fire tube boilers are highly used on all steam locomotives in the horizontal locomotive form. Other than used as locomotives, they are extensively used in stationary engineering fields and producing low pressure steam like for heating of a building.

Operation

The fuel is burnt in a firebox to generate hot combustion gases. Steam is produced by heating water along a series of fire tubes perforating the boiler. The generated steam raises the steam dome, which also controls the exit of steam. In regards to superheat and dry generated steam, saturated steam is always forced into the super heater through the larger flues at the back in the locomotive fire tube boiler. With the help of steam directed to the turbine or cylinders, mechanical power is induced and the exhaust gases are used to heat the feed water for increasing boiler's efficiency.

Advantages

Fire tube boilers are perfectly used for space heating and various industrial process applications. They are compact in size, easy to clean and maintain. The tubes are easily replaceable and cost maintenance is very low. They are available in wide range of sizes.

Disadvantages

- They are limited for generating high capacity steam.
- Fire tube boilers are not appropriate for very high pressure applications.

Types of Fire tube boilers

- Cornish Boiler: This boiler has a long horizontal cylinder with a single large flue consisting the fire.

- Lancashire boiler: Same as Cornish boiler, this has two large flues comprising the fires.
- Scotch marine boiler: This form of boiler uses large number of small diametre tubes. It provides greater heating surface arena for both weight and volume.
- Locomotive boiler: It has three major constituents, double walled firebox, horizontal and cylindrical boiler barrel consists with small flue tubes and smoke box provided with chimney for exhaust gases. They are vastly used in traction engines, steam rollers, portable engines and some other steam road vehicles.

BOILER MOUNTINGS

Being a decade old distributor of WJ Boiler Mountings, we have become the most preferred choice of the market. With the WJ boiler mountings at hand we have an edge over every other steam boiler mountings suppliers of India. The safety boiler mountings that we deal in are of the best quality; leaving behind every speculation about us and of the products. These products are assured to have high durability, thus, proving to be highly advantageous in the long run. These distinct boiler mountings are available in the market at some of the most pocket friendly prices.

It includes:

- Feed check valve
- Accessible feed check valve
- Pressure reducing valve
- Blow off cock
- Steam stop valve
- Non-return valve
- Safety valve
- Steam trap
- Strainer
- Manifold

Safety Flow Control Valves

We are mainly engaged in supplying of safety flow control valves, which includes gate, globe, check, wafer type single/dual plate check, ball, butterfly, diaphragm, self lubricated plug type, knife edge, jacketed sluice, penstock, needle, non-return, flush bottom, balancing, breather etc in different material of construction.

IMPORTANT BOILER MOUNTINGS

A boiler just cant run on its basic working system. There are many external equipments that help a boiler run efficiently. For example, it would have been too difficult to keep a continuous watch on the boiler water level had there not been cascade or boiler automation system. In the same way there are a few important boiler mountings that are fitted on the boiler body itself for a safe and efficient running of the boiler. These fittings are the basic requirements of the whole steam production system. The main boiler mountings are:

Safety Valves

You would always find safety valves mounted in pairs on main boiler body. Safety valves are provided on a boiler to protect the boiler from overpressure. All the safety valves are preset at a particular fixed boiler pressure. This means that the valves are set at a particular lifting pressure and locked to prevent any kind of manipulation later on. The safety valves operate automatically when the boiler reaches the pre-set blow off pressure.

Main Steam Valve

This is the most important valve of the whole system. Main steam stop valve is a non return valve, fitted on the main steam line just near to the boiler. It is the main valve by which the supply of steam from the boiler to all the ship's systems can be started or stopped.

Auxiliary Stem Stop Valve

This is like a back up valve to the main steam stop valve. This means that in case some problem occurs with the main steam stop valve, the auxiliary valve still protects the system from any kind of damage. It is also a non return valve type.

Feed Check or Control Valve

These valves are also found in pairs. One of the valve out of the two valves is the main valve and the other is an auxiliary or stand by valve. These valves are also non return valves and they always carry an indication of their open or closed position.

Feed Check or Control Valve

These valves are also found in pairs. One of the valve out of the two valves is the main valve and the other is an auxiliary or stand by valve. These valves are also non return valves and they always carry an indication of their open or closed position.

Water Level Gauges

Water level gauges are one of the most important mountings of the boiler. Always found in pair, the water level gauges help in maintaining and monitoring the appropriate water level in the boiler. Thus they help in preventing accidents and damage to the boiler. The gauges are always fitted at the opposite ends of the boiler to prevent wrong reading due to motion of the ship. The type of the water gauge fitted depends on the the working pressure of the boiler

Pressure Gauge Connection

Pressure gauges are fitted on the boiler drums and superheaters for pressure readings. They should be periodically checked to prevent accidents due to wrong readings.

Air Release Cock

An important equipment for the boiler starting procedure, the air release cock are fitted in boiler drums, headers etc to purge out the air when filling the boiler or during raising steam

Sampling Connection

This connection is mainly for taking feed water samples for testing and analysis purposes. The samples are collected through this connection and if any kind of chemicals that are to be added to the boiler water they are injected through the this connection.

Blow Down Valve

Used for emptying the boiler water, the blow down valve is fitted at the lowermost end of the boiler. It is also used for partial emptying the boiler and is an important valve in the blow down process.

Scum Valve

This valve is used for removal of impurities from the boiler water. The scum or impurities float on the top of the water. A shallow dish is positioned at the normal water level to collect the scum from the top of the water level.

Whistle Stop valve

It is small bore non return valve which is used to supply steam to the whistle straight from the boiler drum

Additional Mountings on Water Tube boiler

Water tube boilers use less water and produce more steam in comparison to it.Due to this reason the chances of accidents due to high temperature and pressure are greater. Due to this reason, water tube boilers have extra mountings

Automatic Feed Water Regulator

Fitted in the feed water line just near the main check valve, the automatic feed water regulator ensures correct water level in the boiler in spite of the fluctuating load conditions. Boilers having high evaporation rate use multiple element feed water regulator

Low Alarm Level

Produces an alarm sounding when the boiler reaches low water level condition.

High Alarm Level

Produces an alarm sounding when the boiler reaches high water level condition

Super Heater Circulating Valves

Used during the initial warming up process and during the boiler starting procedure, the valves ensures a steady flow of steam and also act as air vents

Sootblowers

Generally used during the boiler maintenance procedure.Sootblowers use pressurized steam or compressed air to blow off the soot and other combustion products deposits from the surface of the tube.

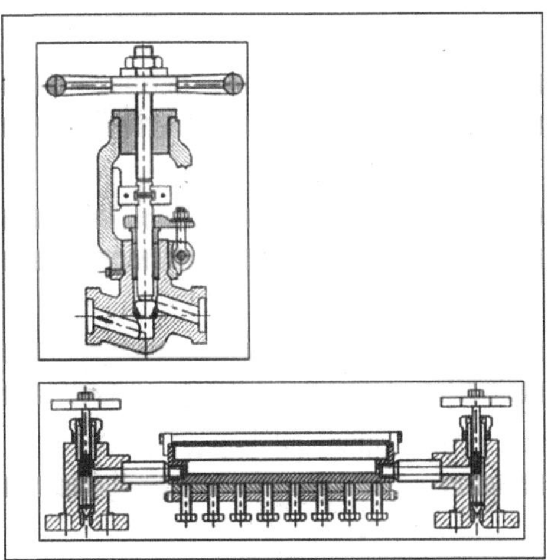

Soot Blow Valves Work On Steam Use To Clean Surfaces In Water Tube Boiler From Deposit Of Fuel Dust Otherwise Rate Of Heat

Transfer Reduces and Boiler Steam Generation and Efficiency Also Reduces.

Types:

- R otary Valve (Manual and Motorised)
- Retractable Long and Short Type.

Gate/Globe Valve, High Pressure Gate/Globe Valves As Up To 63 Kg/cm2. gauge glass assembly.

BOILER COMPONENTS

Our cutting edge technology backed by a pool of experienced professionals aid in the smooth fabrication of our wide assortment of Boiler Components including boiler drums, boiler super heaters, economizer coils, boiler headers, boiler r&w drums, industrial boiler drums, fabricated boiler drums, heaters, super heaters, industrial boiler super heaters, fabricated boiler super heaters, cooling coils, heating coils.

Boiler Headers (Steam Distribution Headers)

The Boiler Headers are fabricated using premium quality raw material and are in high demand in the market. These are also offered in customised form adhering to the designs and specifications of our valued customers and are manufactured in conformation to the international standards.

Boiler Super Heaters

Our Boiler Super Heaters are designed to perfection and are stringently tested as per well-defined parametres. These are meticulously designed by our qualified team of professionals and are widely used all round the globe in different industries and are stringently checked for quality, durability and flawlessness.

Boiler R&W Drums

The Boiler R&W Drums available with us are in high demand all across the globe. Our team of qualified personnel take every care in the production of the best range ever and delivering them in an estimated time frame. Further, we endeavor to satisfy our clients by manufacturing our range as per their requirements.

Economizer Coils

We are engaged in the manufacturing and exporting Economizer Coils that are high in performance and low on maintenance. Designed to perfection these are widely used in different industries all across the world and are fabricated from premium quality raw material conforming to the international standards.

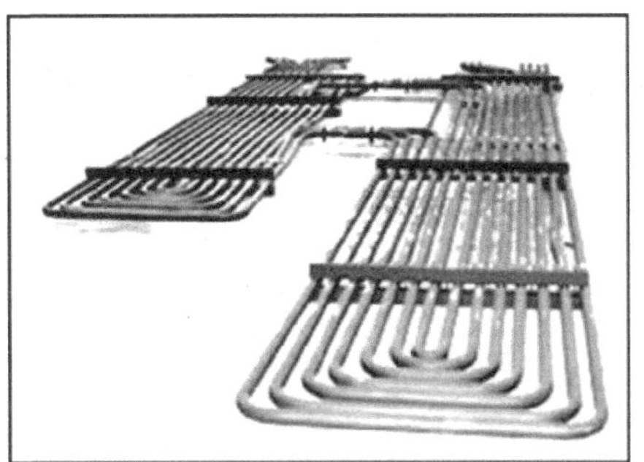

BOILER ACCESSORIES

BOILER INTEGRAL PIPING

Our team of adept professionals intricately designs our Boiler Integral Pipings. They utilize various sophisticated machinery and innovative techniques in the fabrication of our range. In our endeavor to satisfy our clients all across the globe, we also manufacture our gamut in customised form adhering to the designs and specifications of our esteemed clients.

Turbine Integrated Piping

We are available with Turbine Integrated Piping that finds wide application indifferent industries worldwide. These are fabricated using best grades of raw material and are in conformation to the international standards. Our team of quality controllers stringently checks our range for precision, durability, quality and flawlessness.

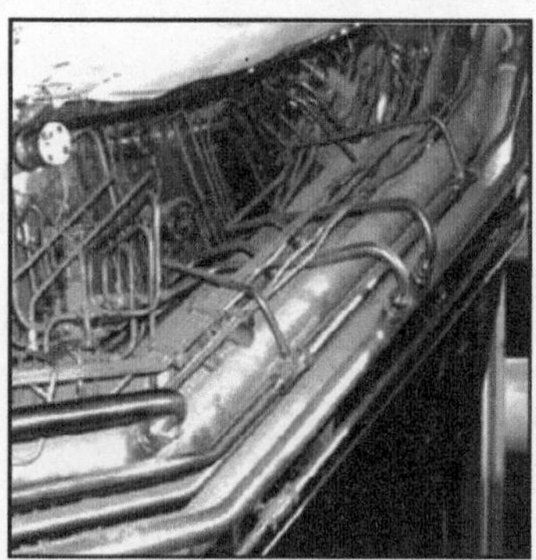

HRSG Integrated Piping

The HRSG Integrated Piping offered by us are of world class designs and are in high demand in the domestic market as well s the foreign front. These are high in performance, low on maintenance and are durable, corrosion resistant and dimensionally accurate.

Risers and Down Comers for Boilers

Over the years we have been successfully catering our broad client base spread all across the globe offering an unmatched quality range of Risers and Down Comers For Boilers. These are intricately designed using sophisticated technologies and hi-tech machines and conform to the international standards.

Balance of Plant Piping

The Balance Of Plant Pipings offered by us are in high demand worldwide. These are meticulously designed and breathe in quality, durability, and flawlessness. With our unparallel gamut, we have fruitfully attained the market credibility.

Cooling Water and Lubrication Piping

We expertise in the smooth fabrication of Cooling Water and Lubrication Piping that are high in performance, low on maintenance, durable and sturdy. We offer our range in customised form adhering to the designs and specifications of our esteemed clients so as to provide maximum satisfaction to them.

Stainless Steel Piping for Desalination and Process Plants

Our Stainless Steel Piping For Desalination and Process Plants are exclusively designed and are in high demand in different industries worldwide. These are fabricated in conformation to the international standards and are stringently tested as per well-defined parametres.

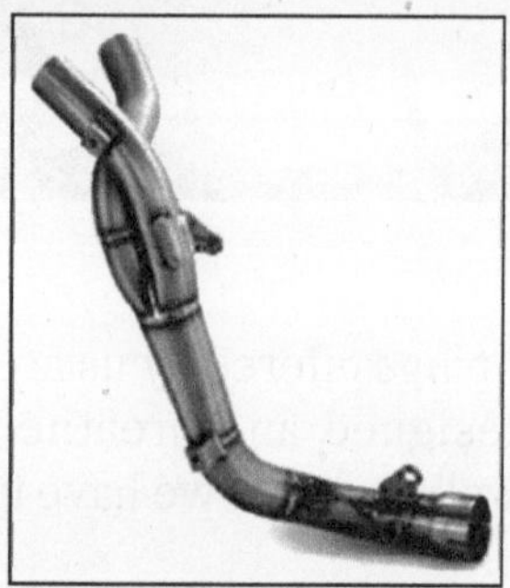

Process Skids

The Process Skids manufactured by us bear world-class designs and are fabricated using best grades of raw material procured from reliable sources. These are highly efficient and are also available in customised form. Besides we also package these equipments as per the requirements of our esteemed clients.

WATER-TUBE BOILERS

A water-tube boiler is one in which the products of combustion (called *flue gas*) pass around tubes containing water. The tubes are interconnected to common water channels and to the steam outlet. For some boilers, baffles to direct the flue gas flow are not required. For others, baffles are installed in the tube bank to direct the flue gas across the heating surfaces and to obtain maximum heat absorption. The baffles may be of refractory or membrane water-tube wall construction, as discussed later. There are a variety of boilers designed to meet specific needs, so care must be exercised in the selection, which should be based on plant requirements, fuel considerations, and space limitations. Water-tube boilers generally may be classified as straight tube and bent tube.

The electric steam boiler provides steam at pressures of about 150 psig. It is a packaged unit generating steam for heating and process. The small units (1000 to 10,000 lb/h of steam) operate at low voltage, while the larger units (7000 to 100,000 lb/h of steam) operate at 13,800 V. Such units find application in educational institutions, office buildings, hospitals, and processing plants. The advantages of the electric steam generator are compactness, no emissions, no stack requirements, quiet operation, low maintenance, safety of operation, absence of fuel storage tanks, the ability to use electric power during off-peak periods, and responsiveness to demand. Its disadvantages lie in the use of high voltage, high power costs, and availability, if it must be used during other than off-peak periods. When such an installation is contemplated, all factors, including the initial cost, need to be considered. It has vertical inclined tubes and has four drums. The upper drums are set on saddles fastened to horizontal beams, the center drum is suspended from an overhead beam by slings, and the lower drum (mud drum) hangs free, suspended from the tubes. Water enters the right-hand drum (top) and flows down the vertical bank of tubes to the lower (mud) drum. It then moves up the inclined bank of tubes because of natural circulation as a result of heating from the flue gases, passing through the center drum and returning to its point of origin, the right-hand drum. Steam passes around a baffle plate.

In the process, most of the entrained moisture is removed before the steam enters the circulators. The steam in passing to the steam drum through the circulators receives a small degree of superheat, approximately 10 to 15°F above the saturation temperature, by the time it enters the steam drum. Superheat is obtained because the circulators are exposed to the flue gas. It is designed with a steep inclination of the main tube bank to provide rapid circulation. Boilers of this general type fit into locations where headroom was a criterion. The tubes are shaped and bent at the ends so that they enter the drum radially.

The upper drums are interconnected by steam circulators (top) and by water circulators (bottom). Note that the center drum has tubes leaving, to enter the rear tube bank. This is done to improve the water circulation. The heating surface is then a combination of waterwall surface, boiler tubes, and a small amount of drum surface. The interdeck superheater and economizer likewise contain a heating surface. The furnace is water-cooled as compared to refractory lined. Downcomers from the upper drum supply water to the sidewall headers, with a steam-water mixture returning to the drum from the wall tubes.

Feedwater enters the economizer, where it is initially heated and then enters the left (top) drum and flows down the rear bank of tubes to the lower (mud) drum. Steam generated in the first two banks of boiler tubes returns to the right and center drums; note the interconnection of drums, top and bottom. Finally, all the steam generated in the boiler and waterwalls reaches the left-hand drum, where the steam is made to pass through baffles or a steam scrubber or a combination of the two, to reduce the moisture content of the steam before it passes to the superheater. The superheater consists of a series of tube loops.

The steam then passes to the main steam line or steam header. The upper drums are supported at the ends by lugs resting on steel columns. The lower drum is suspended from the tubes and is free to move by expansion, imposing no hardship on the setting. The superheater headers are supported (at each end) from supports attached to steel columns overhead. The flue gas resulting from combustion passes over the first bank of tubes, through the superheater, and down across the second pass; the flue gas then reverses in direction and flows up through the third pass. Note the baffles that direct the flow of the flue gas. After leaving the boiler, the flue gas enters the economizer, traveling down (in a counterflow direction to the water flow) through the tubes to the economizer flue gas outlet and then to the stack.. Most of the steam is generated in the waterwalls and the first bank of boiler tubes, since this heating surface is exposed to radiant heat. This unit is fired with a spreader stoker burning coal.

If burners are part of the design, it also can be fired with natural gas, oil, or a combination of the two, providing flexibility in operation. Fly ash (containing unburned carbon particles) is collected at the bottom of the third pass and from the economizer and is reinjected back onto the grate through a series of nozzles located in the rear wall, thus improving the boiler efficiency and lowering fuel costs. This is referred to as *cinder reinjection*. Over-fire air is introduced through nozzles in the rear walls and sidewalls to improve combustion. Although the three- and four-drum designs, as well as the cross-drum Design have been replaced with the modern two-drum and one-drum designs, many of these older units are still in operation today. The increasing cost of field-assembled boilers led to the development of the shop-assembled package boiler for industrial purposes. There are many varieties of packaged boilers. The smaller units are completely factory assembled and ready for shipment. The larger units are of modular construction with final assembly and erection done in the field. The FM package boiler is available in capacities from 10,000 to 200,000 lb/h of steam, with steam pressures of 525 to 1050 psi and steam temperatures to 825°F. Superheaters, economizers, and air preheaters can be added, with operating and economic design consideration. The packaged unit includes burners, soot blowers, forced draft fan, controls, etc. The steam drum is provided with steam separating devices to meet steam purity requirements. The features of the FM boiler are:

- furnace waterwall cooling including sidewalls, rear wall, roof, and floor, which eliminates the need for refractory and its associated maintenance;
- a gas-tight setting preventing gas leaks;
- a steel base frame that supports the entire boiler;
- an outer steel lagging permitting an outdoor installation;
- drum internals providing high steam purity;
- tube-bank access ports providing ease in inspection; and
- soot blowers providing boiler bank cleaning.

The shop-assembled unit can be placed into service quickly after it is set in place by connecting the water, steam, and electric lines; by making the necessary flue connection to the stack; and depending on the size of the unit, by connecting the forced-draft fan and the associated duct. Flue gases flow from the burner to the rear of the furnace (i.e., through the radiant section), reversing to pass through the superheater and convection passes. These units are oil, gas, or combination fired. A combustion control system accompanies the boiler installation. The advantages of package boilers over field-erected boilers are:

- lower cost,
- proven designs,
- shorter installation time, and
- generally a single source of responsibility for the boiler and necessary auxiliaries.

However, package boilers are limited by size because of shipping restrictions and generally can fire only gas and oil. Nevertheless, where shipping clearances allow, package boilers are designed for capacities up to 600,000 lb/h and steam pressures to 1800 psig and steam temperatures to 900°F. These units are oil and gas fired and can be designed for capacities ranging from 60,000 to 200,000 lb/h with steam pressures to 1000 psi and steam temperatures to 800°F. Designs vary between manufacturers, as do the design pressures and temperatures.

The SD steam generator has been developed to meet demands of power and process steam in a wide range of sizes. This unit is available in capacities to 800,000 lb/h of steam, steam pressures to 1600 psi, and steam temperatures to 960°F. It is a complete waterwall furnace construction, with a radiant and convection superheater. If economics dictate, an air preheater or economizer (or combination) can be added. The unit shown is for a pressurized furnace (i.e., incorporating only a forced-draft [FD] fan) designed for oil, gas, or waste fuels such as coke-oven or blast furnace gases. The combination radiant-convection superheaters provide a relatively constant superheat temperature over the normal operating range.

When superheat control is necessary, i.e., a constant superheat temperature over a desired load range, a spray-type desuperheater (attemperator) is installed. This attemperator reduces the higher heat absorption at high loads to ensure that the desired steam temperature is maintained over the load range. This attemperator is located in an intermediate steam temperature zone that ensures mixing and rapid, complete evaporation of the injected water. The rear drum and lower waterwall headers are bottom supported, permitting upward expansion.

Wall construction is such that the unit is pressure-tight for operating either with a balanced draft (using both an FD and an ID fan) or as a pressurized unit. The steam drum is equipped with chevron dryers and horizontal separators to provide dry steam to the superheater. The prominent nose at the top of the furnace ensures gas turbulence and good distribution of gases as they flow through the boiler. The unit is front fired and has a deep furnace, from which the gases pass through the radiant superheater, down through the convection superheater, up and through the first bank of boiler tubes, and then down through the rear bank of tubes to the boiler exit (to the right of the lower drum). The outer walls are of the welded fin-tube type; baffles are constructed of welded fin tubes; the single-pass gas arrangement reduces the erosion in multiple-pass boilers that results from sharp turns in the gas stream.

High-temperature water (HTW) boilers

High-temperature water (HTW) boilers provide hot water under pressure for space heating of large areas such as buildings. Water is circulated at pressures up to 450 psig through the system. The water leaves the HTW boiler at subsaturated temperatures (i.e., below the boiling point at that pressure) up to 430°F. (Note that the boiling temperature at 450 psig is approximately 458°F.) Sizes generally range to about 60 million Btu/h for package units and larger for field-erected units, and the units are designed for oil or natural gas firing.

Most units are shop assembled and shipped as packaged units. The large units are shipped in component assemblies, and it is necessary to install refractory and insulation after the pressure parts are erected. The high-pressure water can be converted into low-pressure steam for process. For example, with a system operating at a maximum temperature of 365°F, the unit is capable of providing steam at 100 psi. A *high-temperature water system* is defined as a fluid system operating at temperatures above 212°F and requiring the application of pressure to keep the water from boiling. Whereas a steam boiler operates at a fixed temperature that is its saturation temperature, a water system, depending on its use, can be varied from an extremely low to a relatively high temperature. The average water temperature within a complete system will vary with load demand, and as a result, an expansion tank is used to provide for expansion and contraction of the water volume as its average temperature varies.

To maintain pressure in the system, steam pressurization or gas pressurization is used in the expansion tank. For the latter, air or nitrogen is used. The pressure is maintained independent of the heating load by means of automatic or manual control. Firing of each boiler is controlled by the water temperature leaving the boiler. The hot-water system is advantageous because of its flexibility. For the normal hot-water system there are no blowdown losses and little or no makeup, installation costs are lower than for a steam-heating system, and the system requires less attention and maintenance. The system can be smaller than an equivalent steam system because of the huge water-storage capacity required by a steam system; peak loads and pickup are likewise minimized, with resulting uniform firing cycles and higher combustion efficiency. The high-temperature water system is a closed system.

When applied to heating systems, the largest advantage is for the heating of multiple buildings. For such applications, the simplicity of the system helps reduce the initial cost. Only a small amount of makeup is required to replace the amount of water that leaks out of the system at valve stems, pump shafts, and similarly packed points. Since there is little or no free oxygen in the system, return-line corrosion is reduced or eliminated, which is in contrast to wet returns from the steam system, wherein excessive maintenance is frequently required. Feedwater treatment can be reduced to a minimum.

Comparison of fire- and water-tube boilers

Fire-tube boilers ranging to 800 bhp (approximately 28,000 lb/h) and oil- and gas-fired water-tube boilers with capacities to 200,000 lb/h are generally shop assembled and shipped as one package. The elimination of field-assembly work, the compact design, and standardization result in a lower cost than that of comparable field-erected boilers. Fire-tube boilers are preferred to water-tube boilers because of their lower initial cost and compactness and the fact that little or no setting is required. They occupy a minimum of floor space.

Tube replacement is also easier on fire-tube boilers because of their accessibility. However, they have the following inherent disadvantages: the water volume is large and the circulation poor, resulting in slow response to changes in steam demand and the capacity, pressure, and steam temperature are limited. Packaged boilers may be of the fire-tube or water-tube variety.

They are usually oil or gas fired. Less time is required to manufacture packaged units; therefore, they can meet shorter project schedules. Completely shop assembled, they can be placed into service quickly. The packaged units are automatic, requiring a minimum of attention, and hence reduce operating costs. In compacting, however, the furnace and heating surfaces are reduced to a minimum, resulting in high heat-transfer rates, with possible overheating and potential increased maintenance and operating difficulties.

Thus caution must be exercised in the selection of packaged units because the tendency toward compactness can be carried too far. Such compactness also makes the units somewhat inaccessible for repairs. Because these units operate at high ratings and high heat transfer, it is important to provide optimal water conditioning at all times; otherwise, overheating and damage to the boiler may result. Water quality is critical for successful boiler operation. Impurities in the water can quickly destroy the boiler and its components. Poor water quality can damage or plug water-level controls and cause unsafe operating conditions.

Water treatment is provided by a variety of equipment: deaerators for the removal of oxygen, water softeners, chemical additives, and boiler blowdown packages. In all cases, boiler water must be analyzed to determine its composition and the type of water treatment that is required. Having good water quality is paramount in all steam power plants in order to have a successfully operating plant. Whether the plant is base loaded or cycling, having the proper water is key to having an economical plant. However, it has been learned that on cycling plants, good water quality is perhaps even more important. If abnormal chemistry conditions (such as excessive oxygen levels, pH excursions, or contaminants) exist during startups, load-following operations, or shutdowns, serious damage can occur, such as corrosion fatigue of the boiler waterwall tubes. The cycling requirements of rapid pressurization and rapid temperature transients, especially during cold startups, can lead to tube failure and thus high temperature costs due to costly repairs and, ultimately, to loss of revenues.

Water-tube boilers are available in various capacities for highpressure and high-temperature steam. The use of tubes of small diameter results in rapid heat transmission, rapid response to steam demands, and high efficiency. Water-tube boilers require elaborate settings, and initial costs are generally higher than those of fire-tube boilers in the range for which such units are most frequently designed. However, when this capacity range is exceeded for highpressure and high-temperature steam, only the water-tube boiler is available.

Air filtration, which plagued the earlier water-tube boiler, has now been minimized in the design by means of membrane tube water walls, improved expansion joints, and casings completely enclosing the unit. Feedwater regulation is no longer a problem when the automatic feedwater regulator is used. In addition, water-tube boilers are capable of burning any economically available fuel with excellent efficiency, whereas packaged units must use liquid or gaseous fuels to avoid fouling the heating surfaces. Thus, in selecting a boiler, many factors other than first cost are to be considered. Important are availability, operating and maintenance costs, fuel costs, space, and a host of other factors. Most important perhaps are fuel costs. During the life of the equipment, we can expect fuel costs to be many times the cost of the boiler and associated equipment.

STEAM-WATER SEPARATION

In the past, difficulty with carry-over and impurities in the steam was frequently encountered (*carry-over* is the passing of water and impurities to the steam outlet). Efforts were directed to reduce carry-over to a minimum by separating the water from the steam through the installation of baffles and the dry pipe. Both measures met with some success. The dry pipe ran the length of the drum, the ends being closed and the upper side of the pipe being drilled with many small holes. The top center of this pipe was connected to the steam outlet.

Steam entering through the series of holes was made to change direction before entering the steam outlet, and in the process the water was separated from the steam. The bottom of the dry pipe contained a drain, which ran below the normal water level in the drum. The dry pipe was installed near the top of the drum so as not to require removal for routine inspection and repairs inside the drum. The dry pipe proved to be fairly effective for small boilers but unsuited for units operating at high steam capacity. Placing a baffle ahead of the dry pipe offered some slight improvement in steam quality but was still not considered entirely satisfactory. Modern practice requires high-purity steam for process, for the superheater, and for the turbine. An important contribution to increased boiler capacity and high rating is the fact that the modern boiler is protected by clean, high-quality feedwater.

The application of both external and internal feedwater treatment is supplemented by the use of steam scrubbers and separators that are located in the steam drum. Steam drums are used on recirculating boilers that operate at subcritical pressures.The primary purpose of the steam drum is to separate the saturated steam from the steam-water mixture that leaves the heattransfer surfaces and enters the drum.

The steam-free water is recirculated within the boiler with the incoming feedwater for further steam generation. The saturated steam is removed from the drum through a series of outlet nozzles, where the steam is used as is or flows to a superheater for further heating. (By definition, saturated steam is pure steam that is at the temperature that corresponds to the boiling temperature at a particular pressure. For example, saturated steam at a pressure of 500 psia has a temperature of 467°F.) The steam drum is also used for the following:

1. To mix the saturated water that remains after steam separation with the incoming feedwater
2. To mix the chemicals that are put into the drum for the purpose of corrosion control and water treatment
3. To purify the steam by removing contaminants and residual moisture
4. To provide the source for a blowdown system where a portion of the water is rejected as a means of controlling the boiler water chemistry and reducing the solids content
5. To provide a storage of water to accommodate any rapid changes in the boiler load.

The most important function of the steam drum, however, remains as the separation of steam and water. Separation by natural gravity can be accomplished with a large steam-water surface inside the drum. This is not the economical choice in today's design because it results in larger steam drums, and therefore the use of mechanical separation devices is the primary choice for separation of steam and water. Efficient steam-water separation is of major importance because it produces high-quality steam that is free of moisture. This leads to the following key factors in efficient boiler operation:

1. It prevents the carry-over of water droplets into the superheater, where thermal damage could result.
2. It minimizes the carry-under of steam with the water that leaves the drum, where this residual steam would reduce the circulation effectiveness of the boiler.
3. It prevents the carry-over of solids. Solids are dissolved in the water droplets that may be entrained in the steam if not separated properly. By proper separation, this prevents the formation of deposits in the superheater and ultimately on the turbine blades.

Boiler water often contains contaminants that are primarily in solution. These contaminants come from impurities in makeup water, treatment chemicals, and leaks within the condensate system such as the cooling water. Impurities also occur from the reaction of boiler water and contaminants with the materials of the boiler and of the equipment prior to entering the boiler. The steam quality of a power plant depends on proper steam-water separation as well as the feedwater quality, and this is a major consideration to having a plant with high availability and low maintenance costs. Even low levels of solids in the steam can damage the superheater and turbine, causing significant outages, high maintenance costs, and loss of production revenues. Prior to the development of quality steam-water separators, gravity alone was used for separation. Because the steam drum diameter requirements increased significantly, the use of a single drum became uneconomical, and therefore it became necessary to use multiple smaller drums. Depending on the size of the boiler, there is a single or double row of cyclone steam separators with scrubbers running the entire length of the drum. Baffle plates are located above each cyclone, and there is a series of corrugated scrubber elements at the entrance to the steam outlet. Water from the scrubber elements drains to a point below the normal water level and is recirculated in the boiler. Operation is as follows:

- The steam-water mixture from the risers enter the drum from behind the baffle plate before entering the cyclone; the cyclone is open at top and bottom.
- Water is thrown to the side of the cyclone by centrifugal force.
- Additional separation of water and steam occurs in the passage of steam through the baffle plates.
- On entering the scrubber elements, water is also removed with steam passing to the steam outlet.

Separators of this type can reduce the solids' carry-over to a very low value depending on the type of feedwater treatment used, the rate of evaporation, and the concentration of solids in the water. The cyclone and scrubber elements are removable for cleaning and inspection and are accessible from manways that are located in the ends of the steam drum. The combination of cyclone separators and scrubbers provides the means for obtaining steam purity corresponding to less than 1.0 part per million (ppm) solids content under a wide variation of operating conditions. This purity is generally adequate in commercial practice; however, the trend to higher pressures and temperatures in steam power plants imposes a severe demand on steam-water separation equipment.

9

The Gas Turbine Plant

GAS TURBINE WORKING PRINCIPLE

Gas turbine engines derive their power from burning fuel in a combustion chamber and using the fast flowing combustion gases to drive a turbine in much the same way as the high pressure steam drives a steam turbine.

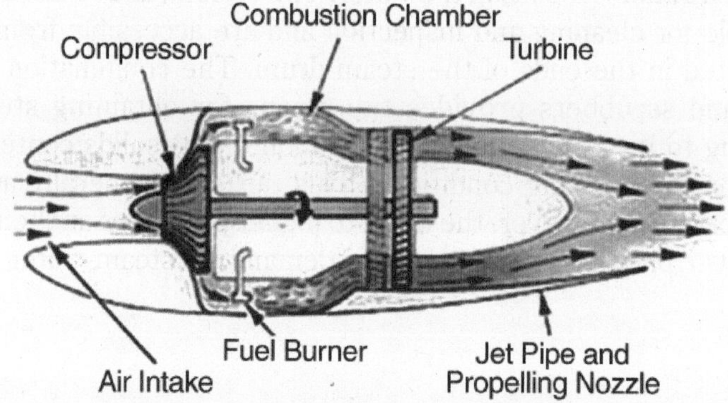

One major difference however is that the gas turbine has a second turbine acting as an air compressor mounted on the same shaft. The air turbine (compressor) draws in air, compresses it and feeds it at high pressure into the combustion chamber increasing the intensity of the burning flame.

It is a positive feedback mechanism. As the gas turbine speeds up, it also causes the compressor to speed up forcing more air through the combustion chamber which in turn increases the burn rate of the fuel sending more high pressure hot gases into the gas turbine increasing its speed even more. Uncontrolled runaway is prevented by controls on the fuel supply line which limit the amount of fuel fed to the turbine thus limiting its speed.

The thermodynamic process used by the gas turbine is known as the Brayton cycle. Analogous to the Carnot cycle in which the efficiency is maximised by increasing the temperature difference of the working fluid between the input and output of the machine, the Brayton cycle efficiency is maximised by increasing the pressure difference across the machine. The gas turbine is comprised of three main components: a compressor, a combustor, and a turbine. The working fluid, air, is compressed in the compressor (adiabatic compression - no heat gain or loss), then mixed with fuel and burned by the combustor under constant pressure conditions in the combustion chamber (constant pressure heat addition).

The resulting hot gas expands through the turbine to perform work (adiabatic expansion). Much of the power produced in the turbine is used to run the compressor and the rest is available to run auxiliary equipment and do useful work. The system is an open system because the air is not reused so that the fourth step in the cycle, cooling the working fluid, is omitted.

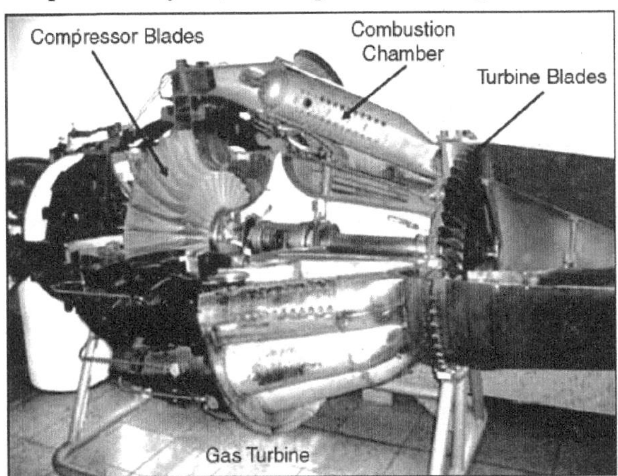

Fig. Gas Turbine Aero Engine (Deutches Museum)

Gas turbines have a very high power to weight ratio and are lighter and smaller than internal combustion engines of the same power. Though they are mechanically simpler than reciprocating engines, their characteristics of high speed and high temperature operation require high precision components and exotic materials making them more expensive to manufacture.

History of Electrical Power Generation

In electricity generating applications the turbine is used to drive a synchronous generator which provides the electrical power output but because the turbine normally operates at very high rotational speeds of 12,000 r.p.m or more it must be connected to the generator through a high ratio reduction gear since the generators run at speeds of 1,000 or 1,200 r.p.m. depending on the AC frequency of the electricity grid.

Turbine Configurations

Gas turbine power generators are used in two basic configurations:

1. Simple Systems consisting of the gas turbine driving an electrical power generator.

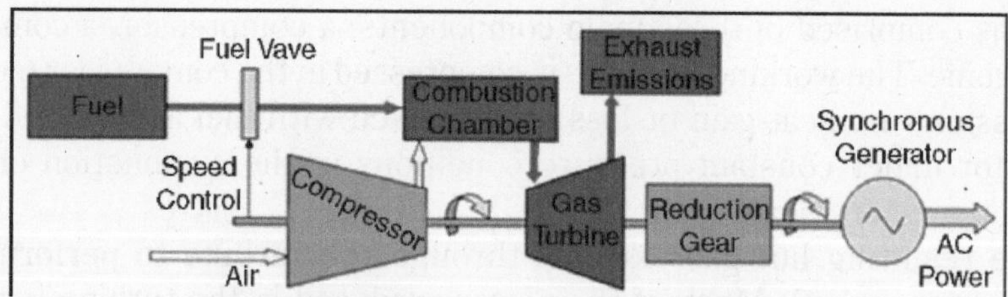

Fig. Gas Turbine Electric Power Generation

2. Combined Cycle Systems which are designed for maximum efficiency in which the hot exhaust gases from the gas turbine are used to raise steam to power a steam turbine with both turbines being connected to electricity generators.

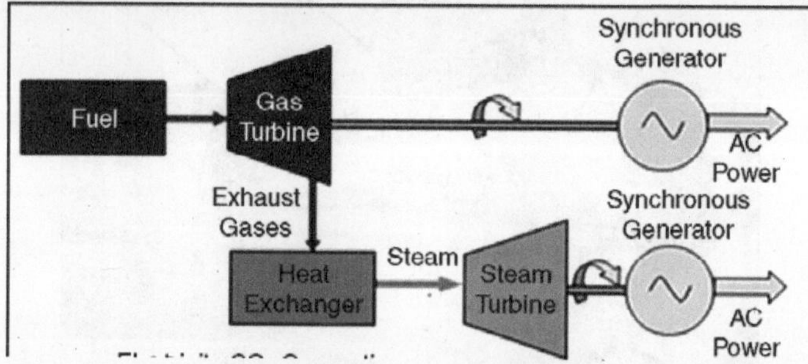

Fig. Electricity Co-generation

Turbine Performance

Turbine Power Output

To minimise the size and weight of the turbine for a given output power, the output per pound of airflow should be maximised. This is obtained by maximising the air flow through the turbine which in turn depends on maximising the pressure ratio between the air inlet and exhaust outlet. The main factor governing this is the pressure ratio across the compressor which can be as high as 40:1 in modern gas turbines.

In simple cycle applications, pressure ratio increases translate into efficiency gains at a given firing temperature, but there is a limit since increasing the pressure ratio means that more energy will be consumed by the compressor.

System Efficiency

Thermal efficiency is important because it directly affects the fuel consumption and operating costs.

Simple Cycle Turbines

A gas turbine consumes considerable amounts of power just to drive its compressor. As with all cyclic heat engines, a higher maximum working temperature in the machine means greater efficiency (Carnot's Law), but in a turbine it also means that more energy is lost as waste heat through the hot exhaust gases whose temperatures are typically well over 1,000°C.

Consequently simple cycle turbine efficiencies are quite low. For heavy plant, design efficiencies range between 30 per cent and 40 per cent. (The efficiencies of aero engines are in the range 38 per cent and 42 per cent while low power microturbines (<100kW) achieve only 18 per cent to 22 per cent).

Although increasing the firing temperature increases the output power at a given pressure ratio, there is also a sacrifice of efficiency due to the increase in losses due to the cooling air required to maintain the turbine components at reasonable working temperatures.

Combined Cycle Turbines

It is however possible to recover energy from the waste heat of simple cycle systems by using the exhaust gases in a hybrid system to raise steam to drive a steam turbine electricity generating set. In such cases the exhaust temperature may be reduced to as low as 140°C enabling efficiencies of up to 60 per cent to be achieved in combined cycle systems. In combined-cycle applications, pressure ratio increases have a less pronounced effect on the efficiency since most of the improvement comes from increases in the Carnot thermal efficiency resulting from increases in the firing temperature. Thus simple cycle efficiency is achieved with high pressure ratios. Combined cycle efficiency is obtained with more modest pressure ratios and greater firing temperatures.

HOW GAS TURBINE POWER PLANTS WORK

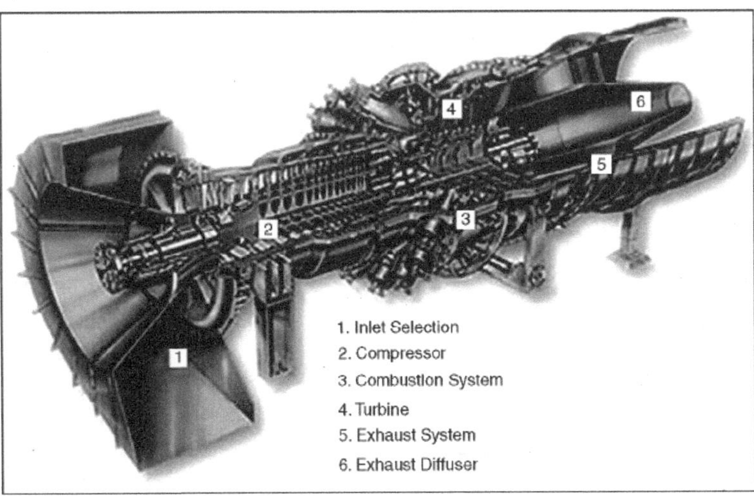

The combustion (gas) turbines being installed in many of today's natural-gas-fuelled power plants are complex machines, but they basically involve three main sections:

1. The compressor, which draws air into the engine, pressurizes it, and feeds it to the combustion chamber at speeds of hundreds of miles per hour.
2. The combustion system, typically made up of a ring of fuel injectors that inject a steady stream of fuel into combustion chambers where it mixes with the air. The mixture is burned at temperatures of more than 2000 degrees F. The combustion produces a high temperature, high pressure gas stream that enters and expands through the turbine section.
3. The turbine is an intricate array of alternate stationary and rotating aerofoil-section blades. As hot combustion gas expands through the turbine, it spins the rotating blades. The rotating blades perform a dual function: they drive the compressor to draw more pressurized air into the combustion section, and they spin a generator to produce electricity.

Land based gas turbines are of two types: (1) heavy frame engines and (2) aeroderivative engines. Heavy frame engines are characterised by lower pressure ratios (typically below 20) and tend to be physically large. Pressure ratio is the ratio of the compressor discharge pressure and the inlet air pressure. Aeroderivative engines are derived from jet engines, as the name implies, and operate at very high compression ratios (typically in excess of 30). Aeroderivative engines tend to be very compact and are useful where smaller power outputs are needed. As large frame turbines have higher power outputs, they can produce larger amounts of emissions, and must be designed to achieve low emissions of pollutants, such as NOx. One key to a turbine's fuel-to-power efficiency is the temperature at which it operates. Higher temperatures generally mean higher efficiencies, which in turn, can lead to more economical operation. Gas flowing through a typical power plant turbine can be as hot as 2300 degrees F, but some of the critical metals in the turbine can withstand temperatures only as hot as 1500 to 1700 degrees F. Therefore, air from the compressor might be used for cooling key turbine components, reducing ultimate thermal efficiency.

One of the major achievements of the Department of Energy's advanced turbine programme was to break through previous limitations on turbine temperatures, using a combination of innovative cooling technologies and advanced materials. The advanced turbines that emerged from the Department's research programme were able to boost turbine inlet temperatures to as high as 2600 degrees F - nearly 300 degrees hotter than in previous turbines, and achieve efficiencies as high as 60 per cent.

Another way to boost efficiency is to install a recuperator or heat recovery steam generator (HRSG) to recover energy from the turbine's exhaust. A recuperator captures waste heat in the turbine exhaust system to preheat the compressor discharge air before it enters the combustion chamber. A HRSG generates steam by capturing heat from the turbine exhaust. These boilers are also known as heat recovery steam generators. High-pressure steam from these boilers can be used to generate additional electric power with steam turbines, a configuration called a combined cycle.

A simple cycle gas turbine can achieve energy conversion efficiencies ranging between 20 and 35 per cent. With the higher temperatures achieved in the Department of Energy's turbine programme, future hydrogen and syngas fired gas turbine combined cycle plants are likely to achieve efficiencies of 60 per cent or more. When waste heat is captured from these systems for heating or industrial purposes, the overall energy cycle efficiency could approach 80 per cent.

POWER PLANTS

DESIRABLE PROPERTIES

Properties of Decomposition

1. Another desirable property in database design is dependency preservation.
   - We would like to check easily that updates to the database do not result in illegal relations being created.
   - It would be nice if our design allowed us to check updates without having to compute natural joins.
   - To know whether joins must be computed, we need to determine what functional dependencies may be
     tested by checking each relation individually.
   - Let F be a set of functional dependencies on schema R.
   - Let {R1,R2...,Rn} be a decomposition of R.
   - The restriction of F to Ri is the set of all functional dependencies in F+ that include only attributes of Ri.
   - Functional dependencies in a restriction can be tested in one relation, as they involve attributes in one
     relation schema.
   - The set of restrictions F1, F2..., Fn is the set of dependencies that can be checked eciently.
   - We need to know whether testing only the restrictions is suffcient.
   - Let F0 = F1, F2..., Fn.
   - F0 is a set of functional dependencies on schema R, but in general, F0
     6= F.

- However, it may be that F'+ = F+.
- If this is so, then every functional dependency in F is implied by F0, and if F0 is satis ed, then F must also be satisfied.
- A decomposition having the property that F0+ = F+ is a dependency-preserving decomposition.

2. The algorithm for testing dependency preservation follows this method.

```
compute F+ for each schema Ri in D do
begin
Fi.= the restriction of F+ to Ri,
end
F' = Ð¤,
for each restriction Fi do
begin
F' = F Fi
end
compute F'+,
if (F'+ = F+) then return (true)
else return (false),
```

3. We can now show that our decomposition of Lending-schema is dependency preserving.
   - The functional Dependency bname → assets bcity can be tested in one relation on Branch-schema.
   - The functional Dependency loan# → amount bname can be tested in Loan-schema.
4. As the above example shows, it is often easier not to apply the algorithm shown to test dependency preservation, as computing F+ takes exponential time.
5. An Easier Way To Test For Dependency Preservation. Really we only need to know whether the functional dependencies in F and not in F0 are implied by those in F'. In other words, are the functional dependencies not easily checkable logically implied by those that are? Rather than compute F+ and F'+, and see whether they are equal, we can do this.
   - Find F - F', the functional dependencies not checkable in one relation.
   - This should take a great deal less work, as we have (usually) just a few functional dependencies to work on.

Desirable Properties of a Database

As you will see, there are many possible choices that can be made during the design and many rules to guide this work. When trying to decide if some choices are better than others, you need to consider the key desirable properties of a database. The table here outlines some of them:

Completeness	Ensures that users can access the data they want. Note that this includes **ad hoc queries**, which would not be explicitly given as part of a statement of data requirements.
Integrity	Ensures that data is both consistent (no contradictory data) and correct (no invalid data), and ensures that users trust the database.
Flexibility	Ensures that a database can evolve (without requiring excessive effort) to satisfy changing user requirements.
Efficiency	Ensures that users do not have unduly long response times when accessing data.
Usability (ease of use)	Ensures that data can be accessed and manipulated in ways which match user requirements.

Developing a 'good' database with these desirable properties isn't easy. Indeed, it is quite possible to develop a database that appears to contain all the relevant data but does not have these properties: as you will see, this has unfortunate consequences.

Exercise: Using the given desirable database properties above, very briefly summarise the characteristics that may be expected with a 'bad' database.

The importance of good database development is found simply in terms of preventing the problems outlined in Solution 2. Database development is not just a matter of creating tables that seem to match the way you see data on forms or reports, but requires a detailed understanding of the meaning of the data and their relationships to ensure that a database has the right properties. This understanding comes from data analysis, which is concerned with representing the meaning of data as a conceptual data model. Getting a conceptual data model right is crucial to the development of a good database.

Desirable Properties of Real-time

Each of these categories is expanded upon below, and later used to compare a number of proposed realtime approaches for Linux. The discussion does go for some time, which is not surprising given that it is summarizing many hundreds of email messages:

• Quality of Service

The traditional view is that the entire operating system is either hard realtime, soft realtime, or non-realtime, but this viewpoint is too coarse grained. Different workloads have different needs, and there is disagreement over the exact definitions of these three categories of realtime. For example, (at least) the following two definitions of "hard realtime" are in use:

- In absence of hardware failures, software provably meets the specified deadlines. This is fine and good, but many applications simply do not need this "diamond hard" realtime.
- Failure to meet the specified deadline results in application failure. This is OK, but -only- if there is a corresponding required probability of success. Otherwise, one could claim "hard realtime" by simply failing the application every time it tries to do anything, which is clearly not useful.

A better approach is to simply specified the required probability of meeting the specified deadline in absence of hardware failure. A probability of 1.0 is consistent with definition (a). Other applications will be satisfied with a probability such as 0.999999, which might be sufficiently high that the probability of software scheduling failure is "in the noise" compared with the probability of hardware failure. A recent LKML thread called this "metal hard" realtime. Or was it "ruby hard"?;-) Of course, one can increase the reliability of hardware through redundancy, but no hardware configuration provides perfect reliability. For example, clusters can increase reliability, so that the probability of failure of the cluster is p ^ n, where "p" is the probability of a single node failing and "n" is the number of nodes. Note that this expression never reaches a probability of 1, no matter how large "n" is. In addition, this mathematical expression assumes that the failover software is perfectly reliable and perfectly configured. This assumption conflicts sharply with my own experience, in which there has always been a point beyond which adding nodes -decreased- cluster reliability.

The timeframe is also critically important. Any system can provide hard realtime guarantees if the deadline is an infinite amount of time in the future. No computer system that I am aware of at this writing is capable of meeting a 1-picosecond scheduling deadline for any task of non-zero duration, but then neither can dedicated digital hardware. Some applications have definite response-time goals, for example, industrial process-control applications tend to have response-time goals ranging from 100s of microseconds to small numbers of seconds. Other applications can benefit from any improvement in response-time goals — faster is better, think in terms of Doom players — but even in these cases there is normally a point of diminishing returns.

The services used by the realtime application also figure in. Given current disk technology, it is not possible to meet a 100-microsecond deadline for a 1MB synchronous write to disk. Not even if you cheat and supply the disk with a battery-backed-up DRAM. However, many realtime applications need only a few of the services that an operating system might provide. This list might include interrupt handling, process scheduling, disk I/O, network I/O, process creation/destruction, VM operations, and so on. Keep in mind that many popular RTOSes provide very little in the way of services! They frequently leave the complex stuff (e.g., web serving) to general-purpose operating systems.

Note that each service can have an associated deadline that it can meet. The interrupt system might be able to meet a 1-microsecond deadline, the real-time process scheduler a 10-microsecond deadline, the disk I/O system a 10-millisecond deadline for moderate-sized I/Os, and so on. The deadline that a service can meet might also depend on the parametres, so that the disk-I/O system would be expected to take longer for larger I/Os.

Furthermore, the probability might vary from service to service or with the parametres to that service. For example, the probability of network I/O completing successfully in minimal time might well be a function of the number of packets transmitted (to account for the probability of packet loss) as well as of packet size (to account for bit-error rate).

To make things even more complicated, the probability of meeting the deadline will vary depending on the length of time allowed. Considering the networking example, a very short deadline might not allow the data transmission to complete, even if it proceeds at wire speed. A longer deadline might allow transmission to complete, but only if there are no transmission errors. An even longer deadline might allow time for a limited number of retransmissions, in order to recover from packet loss due to transmission errors. Of course, a deadline infinitely far into the future would allow guaranteed completion, but I for one am not that patient.

Finally, the performance and scalability of both realtime and non-realtime applications running on the system can be important. Given the current state of the art, one must pay a performance penalty for realtime support, but the smaller the penalty, the better.

So, to sum up, here are the components of a quality-of-service metric for realtime OSes:

- List of services for which realtime response is supported.
- For each service:
  - Probability of missing a deadline due to software, ranging from 0 to 1, with the value of 1 corresponding to the hardest possible hard realtime.
  - Allowable deadline, measured from the time that the request is initiated to the time by which the response must be received.
  - Performance and scalability provided to both realtime and non-realtime applications.
- Amount of Code Inspection Required

So you add a new feature to a realtime operating system. How much of the rest of the system must you inspect and understand in order to be able to guarantee that your new feature provides the required level of realtime response? The smaller this amount of code, the easier it is to add new features and fix bugs, and the greater the number of people who will be able to contribute to the project. In addition, the smaller the amount of such code, the smaller the probability that some well-intentioned bug fix will break realtime response.

Each of the following categories of code might need to be inspected:

- The low-level interrupt-handing code.
  - The realtime process scheduler.
  - Any code that disables interrupts.
  - Any code that disables preemption.
  - Any code that holds a lock, mutex, semaphore, or other resource that is needed by the code implementing your new feature.

Of course, use of automated tools could make such inspection much more reliable and less onerous, but such tools would need to deal with the very large number of CPU architectures and configuration options that Linux supports. The smaller the amount of code that must be inspected, the less chance there is that such a tool will fall victim to configuration-architecture combinatorial explosion.

Each of Linux realtime approaches uses a different strategy to minimize the amount of code in these categories. These differences are surprisingly important, and will be discussed in more detail when going over the various approaches to Linux realtime.

- *API Provided*: I never have learned to -really- like the POSIX API, with the gets() primitive being a particular cause of heartburn, but given the huge amount of software out there that relies on it and the equally huge number of developers who are familiar with it, one should certainly strive to provide it, or at least a sizeable subset of it.

Other popular APIs include the various Java runtime environments, and of course the feared and loathed, but quite ubiquitous, Windows API.

There are a lot of developers and a lot of software out there. The more of these existing developers and software your API supports, the more successful your realtime facility is likely to be.

- *Relative Complexity*: How much realtime capability should be added to the operating system? How much of this burden should the applications take on? Is it better to push some of the complexity into a nanokernel, hypervisor, or other software or firmware layer? Let's first look at the tradeoff between OS and application.

For example, although it is certainly possible to program for separate realtime and non-realtime operating-system instances, doing so adds complexity to the application. Complexity is particularly deadly in the hard realtime arena, and can be literally so if human lives are at risk. Balancing this consideration is the need for simplicity in the operating-system kernel. This balancing act must be carefully considered, taking both the relative complexities and the number of uses into account. Some would argue that it is worthwhile adding 1,000 lines to the OS if that saves 100 lines in each of 1,000 applications. Others would disagree, perhaps citing the greater fault isolation that might be provided by the separation. But this balance clearly must be struck somewhere between writing the application to bare metal on the one hand (but achieving a perfectly simple zero-size operating system) and bloating the operating system beyond the limits of maintainability on the other hand.

Similar arguments can be made for moving some functionality into a hypervisor or nanokernel layer, though fault isolation also comes into play here. Many of the most vociferous arguments seem to revolve around this complexity issue.

- *Fault Isolation*: Can a programming error in a non-realtime application or in a non-realtime portion of the OS harm a realtime application?

Some applications do not care: in these cases, a failure anywhere causes a user-visible failure, so it is not important to isolate faults. Of course, even in these cases, it may be valuable to isolate faults in order to aid debugging, but, other than that, the fault isolation does not help overall application reliability.

In other cases, the realtime portion of the application is protecting someone's life and limb, but the non-realtime portion is only compiling statistics and reports. In this case, fault isolation can be of the utmost importance.

What sorts of faults need isolating?

- Excessive disabling of interrupts
- Excessive disabling of preemption
- Holding a lock, mutex, or semaphore for too long, when that resource must be acquired by realtime code
- Memory corruption, either via wild pointers or via wild DMA

These faults might occur in the main kernel, in a loadable module, or in some debugging tool, such as a kprobe procedure or a kernel-debugger breakpoint script. Though in the latter case, perhaps realtime deadlines should not be guaranteed when actively debugging. After all, straightforward debugging techniques, such as use of kprint(), can cause response-time problems even in non-realtime environments.

- Hardware and Software Configurations

Is SMP required? If so, how many CPUs? How many tasks? How many disks? How many HBAs?

If all the code in the kernel were O(1), it might not matter, but the Linux kernel has not yet reached this goal. Therefore, some applications may choose to restrict the software or the hardware configuration of the platform in order to meet the realtime deadlines. This approach is consistent with traditional

RTOS methodology — RTOS vendors have been known to restrict the configurations in which they will support hard realtime guarantees.

USED FOR POWER PLANT

Three types of low toxicity gas-oil burners have been developed at the firm JSC "NPO TsKTI." Thus far, they have been introduced at a number of electric power stations in this country for converting coal dust burners to combustion of natural gas and oil and for modernizing gas-oil burners, making use of well known in-boiler measures for suppressing the formation of nitrogen oxides. In this way the engineering-economic and ecological performance of the boilers has been improved significantly.

Our renewable energy engineering and renewable energy project development services include: Carbon Credits and Carbon Emissions Consulting, Design, Engineering, Environmental, Feasibility Studies, Feedstock, Legal, Onsite Power Generation (cogeneration or trigeneration) & Greenhouse Gas Emissions consulting for projects located in the U.S. and Canada.

A Circulating Fluidized Bed Boiler is a fully contained state-of-the-art technology for processing solid fuels where fuel is suspended in a mixture of superheated air and sand, collectively called the "fluid bed." Reagents like limestone are added, and temperatures are controlled to directly capture the sulfur and reduce formation of Nitrogen Oxides.

Fluidized Bed Boilers, a Bed for Burning Coal

It was a wet, chilly day in Washington DC in 1979 when a few scientists and engineers joined with government and college officials on the campus of Georgetown University to celebrate the completion of one of the world's most advanced coal combustors. It was a small coal burner by today's standards, but large enough to provide heat and steam for much of the university campus. But the new boiler built beside the campus tennis courts was unlike most other boilers in the world.

A Fluidized Bed Boiler

In a fluidized bed boiler, upward blowing jets of air suspend burning coal, allowing it to mix with limestone that absorbs sulfur pollutants.

It was called a "fluidized bed boiler."

In a typical pulverized coal boiler, coal is crushed into very fine particles, blown into the boiler, and ignited to form a long, lazy flame. In other types of boilers, the burning coal simply rests on grates. But in a "fluidized bed boiler," crushed coal particles "float" inside the boiler, suspended on upward-blowing jets of air. The red-hot mass of floating coal — called the "bed" — would bubble and tumble around like boiling lava inside a volcano. Scientists call this being "fluidized." That's how the name "fluidized bed boiler" came about.

Fluidized Bed Boiler asBurn Coal Cleaner

There are two major reasons fluidized bed boilers are cleaner, and superior to typical coal fired power plants. One, the tumbling action allows limestone to be mixed in with the coal. Remember - limestone is a "sulfur sponge" in that it absorbs sulfur pollutants. As coal burns in a fluidized bed boiler, it releases sulfur. But just as rapidly, the limestone tumbling around beside the coal captures the sulfur. A chemical reaction occurs, and the sulfur gases are changed into a dry powder that can be removed from the boiler.

The second reason a fluidized bed boiler burns cleaner is that it burns "cooler." Cooler in this sense as it is still fairly hot at about 1400 degrees F. But older coal boilers operate at temperatures nearly twice that (almost 3000 degrees F). Also, recall that nitrogen oxides form when a fuel burns hot enough to break apart the nitrogen molecules in the air and cause the nitrogen atoms to join with oxygen atoms. But 1,400 degrees isn't hot enough for that to happen, so few nitrogen oxides forms in a fluidized bed boiler. The result is that a fluidized bed boiler can burn very dirty coal and remove 90% or more of the sulfur and nitrogen pollutants while the coal is burning. Fluidized bed boilers can also burn just about anything else - all types of biomass, including wood, ground-up railroad ties, even soggy coffee grounds.

Today, fluidized bed boilers are operating or being built that are 10 to 20 times larger than the small unit built almost 20 years ago at Georgetown University. There are more than 300 of these boilers operating here in the USA and around the world.

A new type of fluidized bed boiler makes a major improvement in the basic fluidized bed boiler technology. It encases the entire boiler inside a large pressure vessel, much like the pressure cooker used in homes for canning fruits and vegetables — except the ones used in power plants are the size of a small house! Burning coal in a "pressurized fluidized bed boiler" produces a high-pressure stream of combustion gases that can spin a gas turbine to make electricity, then boil water for a steam turbine — two sources of electricity from the same fuel input - that is called "cogeneration."

A "pressurized fluidized bed boiler" is a more efficient way to burn coal. In fact, future boilers using this system will be able to generate 50% more electricity from coal than a regular power plant from the same amount of coal. That's like getting 3 units of power when you used to get only 2. Because it uses less fuel to produce the same amount of power, a more efficient "pressurized fluidized bed boiler" will reduce the amount of carbon dioxide (a greenhouse gas) released from coal-burning power plants.

Coal Gasification

One of the most advanced - and cleanest - coal power plants in the world is Tampa Electric's Polk Power Station in Florida. Rather than burning coal, it turns coal into a gas that can be cleaned of almost all pollutants. This technology is called coal gasification. How do you break apart the atoms of coal? You may think it would take a sledgehammer, but actually all it takes is water and heat. Heat coal hot enough inside a big metal vessel, blast it with steam (the water), and it breaks apart. Into what?

The carbon atoms join with oxygen that is in the air (or pure oxygen can be injected into the vessel). The hydrogen atoms join with each other. A major reason is that the impurities in coal — like sulfur, nitrogen and many other trace elements — can remove practically all of the pollutants when coal is changed into Synthesis Gas through Coal Gasification. In fact, scientists have ways to remove 99.9% of the sulfur and small dirt particles from the coal gas. Coal Gasification is one of the best ways to clean pollutants out of coal.

Another reason is that the coal gases — carbon monoxide and hydrogen, or simply "Synthesis Gas" — don't have to be burned. They can also be used as valuable chemicals. Scientists have developed chemical reactions that turn carbon monoxide and hydrogen into everything from liquid fuels for cars and trucks to plastic toothbrushes! Today, in Tampa, Florida, and West Terre Haute, Indiana, there are power plants generating electricity through "coal gasification" instead of burning it. At a plant in Kingsport, Tennessee, coal gas is being used to make plastic for photographic film and to make methanol (a fuel that can be burned in automobile engines).

Coal Gasification could be one of the most promising ways to use coal in the future to generate electricity and other valuable products. Yet, it is only one of an entirely new family of energy processes called "Clean Coal Technologies" — technologies that can make fossil fuels future fuels.

Fluidized Bed Combustion

Fluidized beds suspend solid fuels on upward-blowing jets of air during the combustion process. The result is a turbulent mixing of gas and solids. The tumbling action, much like a bubbling fluid, provides more effective chemical reactions and heat transfer. Fluidized bed combustion evolved from efforts to find a combustion process able to control pollutant emissions without external emission controls (such as scrubbers). The technology burns fuel at temperatures of 1,400 to 1,700 degrees F, well below the threshold where Nitrogen Oxides form (at approximately 2,500 degrees F, the nitrogen and oxygen atoms in the combustion air combine to form nitrogen oxide pollutants).

The mixing action of the fluidized bed results brings the flue gases into contact with a sulfur-absorbing chemical, such as limestone or dolomite. More than 95 per cent of the sulfur pollutants in coal can be captured inside the boiler by the sorbent. Pressurized Fluidized bed combustion (PFBC) builds on earlier work in atmospheric fluidized-bed combustion technology.

Atmospheric fluidized bed combustion is crossing over the commercial threshold, with most boiler manufacturers currently offering fluidized bed boilers as a standard package. This success is largely due to the Clean Coal Technology Program and the Energy Department's Fossil Energy and industry partners' R&D.

The popularity of fluidized bed combustion is due largely to the technology's fuel flexibility - almost any combustible material, from coal to municipal waste, can be burned - and the capability of meeting sulfur dioxide and nitrogen oxide emission standards without the need for expensive add-on controls. The Clean Coal Technology Program led to the initial market entry of 1st generation pressurized fluidized bed technology, with an estimated 1000 megawatts of capacity installed worldwide. These systems pressurize the fluidized bed to generate sufficient flue gas energy to drive a gas turbine and operate it in a combined-cycle.

The 1st generation pressurized fluidized bed combustor uses a "bubbling-bed" technology. A relatively stationary fluidized bed is established in the boiler using low air velocities to fluidize the material, and a heat exchanger (boiler tube bundle) immersed in the bed to generate steam. Cyclone separators are used to remove particulate matter from the flue gas prior to entering a gas turbine, which is designed to accept a moderate amount of particulate matter (i.e., "ruggedized").

A 2nd generation pressurized fluidized bed combustor uses "circulating fluidized-bed" technology and a number of efficiency enhancement measures. Circulating fluidized-bed technology has the potential to improve operational characteristics by using higher air flows to entrain and move the bed material, and re-circulating nearly all the bed material with adjacent high-volume, hot cyclone separators. The relatively clean flue gas goes on to the heat exchanger. This approach theoretically simplifies feed design, extends the contact between sorbent and flue gas, reduces likelihood of heat exchanger tube erosion, and improves SO2 capture and combustion efficiency.

A major efficiency enhancing measure for 2nd generation pressurized fluidized bed combustor is the integration of coal gasification to produce Synthesis Gas. This fuel gas is combusted in a topping combustor and adds to the combustor's flue gas energy entering the gas turbine, which is the more efficient portion of the combined cycle. The topping combustor must exhibit flame stability in combusting low-Btu gas and low-NOx emission characteristics. To take maximum advantage of the increasingly efficient commercial gas turbines, the high-energy gas leaving the topping combustor must be nearly free of particulate matter and alkali/sulfur content. Also, releases to the environment from the pressurized fluid bed combustion system must be essentially free of mercury, a soon-to-be regulated hazardous air pollutant. To reduce cost and carbon dioxide emissions, new sorbents are being evaluated.

Sorbent utilization has a major influence on operating costs, and carbon dioxide emissions streams can result in the production and use of alkali-based sorbents.

Efforts are ongoing at the Power Systems Development Facility (PSDF) in Wilsonville, Alabama to ensure critical components and subsystems are ready for demonstration of 2nd generation pressurized fluidized bed combustion. The PSDF is operated by Southern Company Services under DOE contract to conduct cooperative R&D with industry.

Tests conducted at the PSDF in 1998 verified that a newly developed multi-annular swirl burner (MASB) provided the needed flame stability and low-NOx performance characteristics. Tests of promising new hot gas filter components and systems are continuing at the PSDF. Advances made to date in this critical technology area include the development of clay-bonded silicon carbide candle filters and the associated filter vessel. Efforts are currently focused on improved candle filter materials for enhanced durability under extreme temperatures and corrosive environment. New ceramics and ceramic-metallic composites are showing promise. Those passing laboratory screening tests will undergo testing at the PSDF.

WORKING FLUID—MERCURY

Mercury has got to be the ultimate dodgy working fluid mercury: as remarked elsewhere on this site, in the Steamwheel gallery, (where mercury is used as a weight and sealant rather than a working fluid) it is an insidious poison of a most unpleasant kind. It has a much higher boiling point than water, at 357 degC. Experimental power station installations were tested in the USA but ultimately nothing came of them. Probably a good thing.

ERCURY IN THE NINETEENTH CENTURY

There was another kind of marine engine that I think should not be passed over without notice; I allude to Howard's quicksilver engine. The experiments with this engine were persevered in for some considerable time, and it was actually used for practical purposes in propelling a passenger steam-vessel called the Vesta, and running between London and Ramsgate. In that engine the boiler had a double bottom, containing an amalgam of quicksilver and lead.

This amalgam served as a reservoir of heat, which it took up from the fire below the double-bottom, and gave forth at intervals to the water above it. There was no water in the boiler, in the ordinary sense of the term, but when steam was wanted to start the engine, a small quantity of water was injected by means of a hand-pump, and after the engine was started, there was pumped by it into the boiler, at each half revolution, as much water as would make the steam needed. This water was flashed on the top surface of the reservoir in which the amalgam was confined, and was entirely turned into steam, the object of the engineers in charge being to send in so much water as would just generate the steam, but so as not to leave any water in the boiler. The engines of the Vesta were made by Mr Penn, for Mr Howard, of the King and Queen Ironworks, Rotherhithe. Mr Howard was, I fear, a considerable loser by his meritorious efforts to improve the steam-engine."

"There was used, with this engine, an almost unknown mode of obtaining fresh water for the boiler. Fresh water, it will be seen was a necessity in this mode of evaporation. The presence of salt, or of any other impurity, when the whole of the water was flashed into steam, must have caused a deposit on the top of the amalgam chamber at each operation.

Fresh water, therefore, was needed; the problem arose how to get it; and that problem was solved, not by the use of surface condensation, but by the employment of reinjection, that is to say, the water delivered from the hot well was passed into pipes external to the vessel; after traversing them, it came back into the injection tank sufficiently cooled to be used again. The boilers were worked by coke fires, urged by a fan blast in their ashpits, but I am not aware that this mode of firing was a needful part of the system."

The idea of using an amalgam of mercury and lead as a reservoir of heat was not a good one, and one regrets that Mr Howard was a considerable loser. The liquid with the highest specific heat, and so greatest storage capacity, is not some exotic chemical. Remarkably, it is plain water. Much cheaper and much safer!

MERCURY IN THE TWENTIETH CENTURY

The man behind the use of mercury vapour turbines in electric power stations was William Le Roy Emmet of the General Electric Company. Emmet devoted a great deal of time and energy to their development and promotion, "as a more efficient power generation system than steam turbines." The difficulties and risks in using the new technology led to termination of the development after his retirement. He died in Erie, Pennsylvania, on September 26, 1941, at the age of 82, so he had presumably managed to avoid mercury poisoning. Mercury has a much higher boiling point than water, so the Carnot efficiency of a thermodynamic cycle that uses it is higher, in theory at least. The plants were dual-cycle, (also known as binary cycle) the hot metallic exhaust from the mercury turbines being used to produce steam for a second stage of steam turbines.

The mercury is vapourised in the boiler, drives the mercury turbine, and is condensed by boiling water to steam. This steam is superheated in the mercury boiler and then drives a conventional steam turbine. The "bled steam heaters" refer to the practice of heating the feed water by bleeding steam from various points along the turbine; this increases cycle efficiency and had been a standard feature of steam power plants for many years.

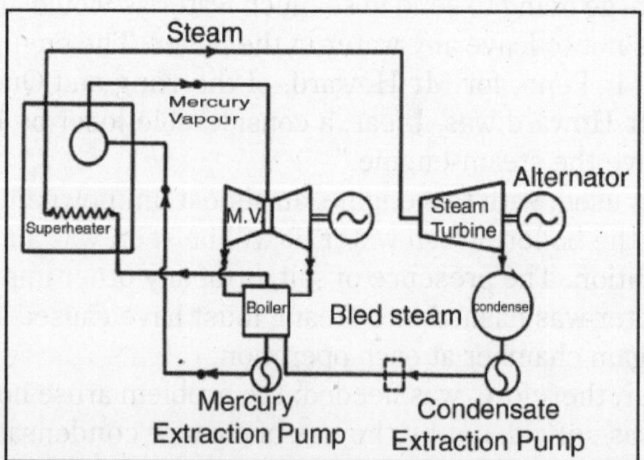

On close inspection this diagram shows exactly the same plumbing as the one above.

According to this book, the mercury circuit operated at 200psi and 538 degC. Electrically welded joints were used in the pipework because the poisonous nature of mercury was recognised, and leaks were not acceptable. Even so, it is hard to believe that a big installation could have been completely free from leaks, and I for one would not have cared to work there, or indeed within ten miles of it.

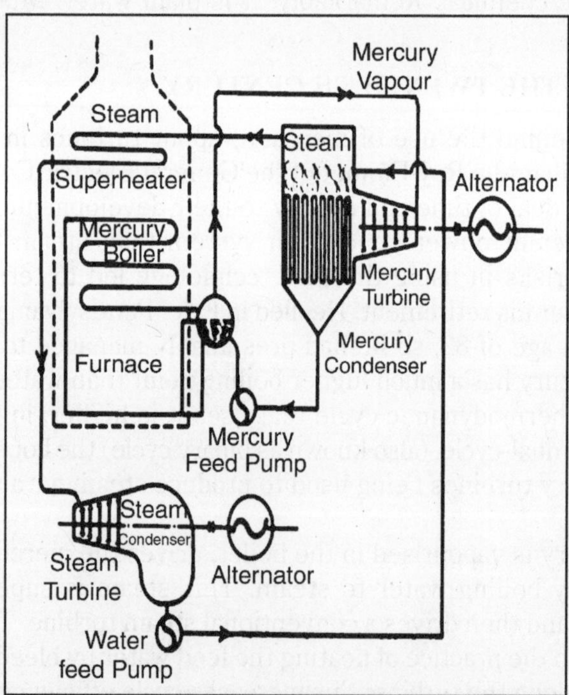

In one American plant (it is not clear which) the mercury circuit operated at 113 psi and 507 degC, driving a 15 MW turbine-alternator set at 720 rpm.

The mercury condenser produced steam at 400 psi and 371 degC, at the rate of 200,000 lb/hour; this drove conventional steam turbines and alternators.

Efficient Use of Steam

"The Efficient Use of Steam" by Oliver Lyle was the bible of steam usage in Britain during the second world war and after. It was aimed more towards process steam rather than power generation, on mercury/water binary cycles: "Mercury boils at 900F under the very modest pressure of 80psi. At 28.5 in vacuum mercury vapour has a temperature of 437 degF (225 degC).

If therefore, mercury is boiled in a boiler, and the saturated mercury vapour passed through a mercury turbine the mercury vapour can be exhausted into a condenser which can be a water boiler, and can raise steam to 250psi.This steam can be given some superheat up to say 600 degF (316 deg C) from the flue gases of the mercury boiler. Owing to the low latent heat of mercury it is necessary to use about 10 pounds of mercury for every 1 pound of water." and later in the text:

"The disadvantages are somewhat formidable. Mercury vapour is extremely poisonous. Mercury does not wet metal surfaces.* The plant is complicated and costly. But several large mercury-steam stations have been working for some years in the USA and have shown very high sustained overall thermal efficiencies. other fluids might give better results. Diphenyl oxide has been suggested as being more suitable than mercury."

If the surfaces are not wetted by the working fluid, then heat transfer would be greatly impaired. But you will see below that mercury *does* wet tantalum. Now that's an interesting point above about the amount of mercury required. And it's not cheap stuff.

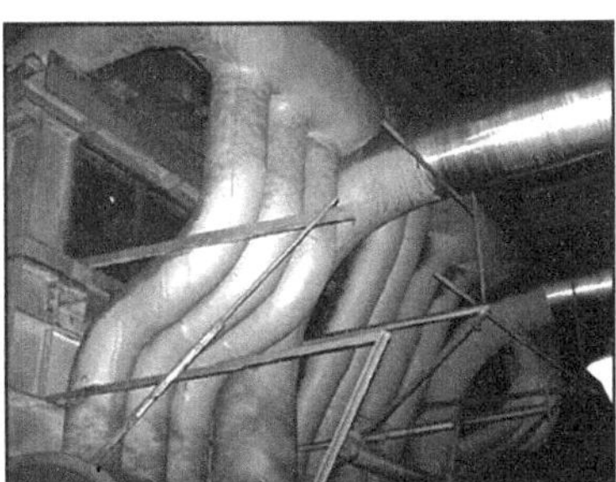

Mercury Power in Space

The reactor is cooled with a mixture of sodium and potassium, (NaK) which heats a mercury boiler. The mercury vapour drives a turbo-alternator and is then condensed and subcooled by a secondary NaK heat rejection loop which transfers the waste heat to a radiator for rejection to space.

The report concentrates on the design of the potassium/mercury boiler, which is an exotic device indeed. It is a spiral stucture designed to fit in a small toroidal space in a cylindrical spacecraft; the heat exchanger tubes are of tantalum, which is readily wetted by mercury, making for good heat transfer. The net power output was 37 kW.

Note the extra cooling and lubricating loop using polyphenyl ether.

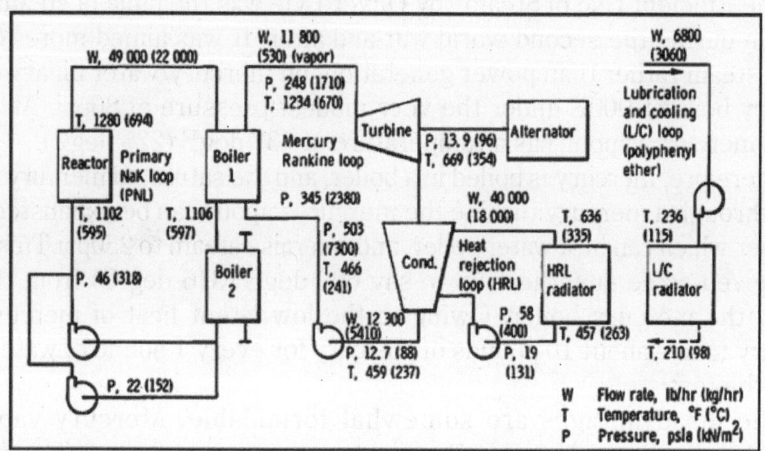

The extraordinary thing is that the overall efficiency seems horribly low at 6.9%. A conventional terrestrial steam power station would give some where around 40%.

So why is it so bad? The Carnot efficiency E = (T1 - T2)/T1 where T1 is the heat input (top) temperature and T2 the heat rejection temperature. The effective rejection temperature in the condenser is a bit less clear but the NaK coolant leaves it at 335 degC. This gives a Carnot efficiency of 35%. That's obviously a theoretical maximum, but a long step from 6.9%; there must be some serious losses somewhere.

The rejection temperature has to be high because in space there are no rivers flowing by to cool your condensers- the only way to lose heat is to radiate it away from a flat plate, which needs to be pretty hot to radiate effectively. Hence the radiators in the flow diagram; these are nothing like central-heating "radiators" which work mostly by convection.

This is one reason- possibly the most important- for the choice of mercury as a working fluid; it permits a high heat rejection temperature so the radiators are efficient, and therefore relatively small and light. This project was only one of several nuclear-powered sytems that were intended to provide large quantities of electricity for long-duration space missions. One of the reasons that they have not so far been used is that no-one is very comfortable with the notion of launching a nuclear reactor on top of a rocket.

INDUSTRIAL AND SMALL POWER PLANTS

Various industries require steam to meet many of their needs: heating and air conditioning; turbine drives for pumps, blowers, or compressors; drying and other processes; water heating; cooking; and cleaning. This so-called industrial steam, because of its lower pressure and temperature as compared with utility requirements, also can be used to generate electricity. This can be done directly with a turbine for electric production only, or as part of a cogeneration system, where a turbine is used for electric production and low-pressure steam is extracted from the turbine and used for heating or for some process. The electricity that is produced is used for in-plant requirements, with the excess often sold to a local electric utility.

Another method is a combined cycle system, where a gas turbine is used to generate electric power and a heat recovery system is added using the exhaust gas from the gas turbine as a heat source. The generated steam flows to a steam turbine for additional electric generation, and this cogeneration results in an improvement in the overall efficiency. The steam that is generated also can be used as process steam either directly or when extracted from the system, such as an extraction point within the turbine. One of the most distinguishable features of most industrial-type boilers is a large saturated water boiler bank between the steam drum and the lower drum. This particular unit is designed to burn pulverized coal or fuel oil, and it generates 885,000 lb/h of steam. Although not shown, this boiler also requires environmental control equipment to collect particulates and acid gases contained in the flue gas. The boiler bank serves the purpose of preheating the inlet feedwater to the saturated temperature and then evaporating the water while cooling the flue gas. In lower-pressure boilers, the heating surface that is available in the furnace enclosure is insufficient to absorb all the heat energy that is needed to accomplish this function.

Therefore, a boiler bank is added after the furnace and superheater, if one is required, to provide the necessary heat-transfer surface. As the pressure increases, the amount of heat absorption that is required to evaporate water declines rapidly, and the heat absorption for water preheating and superheating steam increases. It is also common for boilers to be designed with an economizer and/or an air heater located downstream of the boiler bank in order to reduce the flue gas temperature and to provide an efficient boiler cycle. It is generally not economical to distribute steam through long steam lines at pressures below 150 psig because, in order to minimize the pressure drop that is caused by friction in the line, pipe sizes must increase with the associated cost increase.

In addition, for the effective operation of auxiliary equipment such as sootblowers and turbine drives on pumps, boilers should operate at a minimum pressure of 125 psig. Therefore, few plants of any size operate below this steam pressure. If the pressure is required to be lower, it is common to use pressure-reducing stations at these locations. For an industrial facility where both electric power and steam for heating or a process are required, a study must be made to evaluate the most economical choice. For example, electric power could be purchased from the local utility and a boiler could be installed to meet the heating or process needs only. By comparison, a plant could be installed where both electricity and steam are produced from the same system.

FLUIDIZED BED BOILERS

There are various ways of burning solid fuels, the most common of which are in pulverized-coal-fired units and stoker-fired units. These designs for boilers in the industrial size range have been in operation for many years and remain an important part of the industrial boiler base for the burning of solid fuels. These types of boilers and their features continue to be described in this book. Although having been operational for nearly 40 years, but not with any overall general acceptance, the fluidized bed boiler is becoming more popular in modern power plants because of its ability to handle hard-to-burn fuels with low emissions.

As a result, this unique design can be found in many industrial boiler applications and in small utility power plants, especially those operated by independent power producers (IPPs). Because of this popularity, this book includes the features of some of the many designs available and the operating characteristics of each. In fluidized bed combustion, fuel is burned in a bed of hot particles that are suspended by an upward flow of fluidizing gas. The fuel is generally a solid fuel such as coal, wood chips, etc. The fluidizing gas is a combination of the combustion air and the flue gas products of combustion. When sulfur capture is not required, the fuel ash may be supplemented by an inert material such as sand to maintain the bed. In applications where sulfur capture is required, limestone is used as the sorbent, and it forms a portion of the bed.

Bed temperature is maintained between 1550 and 1650°F by the use of a heat-absorbing surface within or enclosing the bed. As stated previously, fluidized bed boilers feature a unique concept of burning solid fuel in a bed of particles to control the combustion process, and the process controls the emissions of sulfur dioxide (SO2) and nitrogen oxides (NO*x*). These designs offer versatility for the burning of a wide variety of fuels, including many that are too poor in quality for use in conventional firing systems. The state of fluidization in a fluidized bed boiler depends mainly on the bed particle diameter and the fluidizing velocity. There are two basic fluid bed combustion systems, the bubbling fluid bed (BFB) and the circulating fluid bed (CFB), and each operates in a different state of fluidization. At relatively low velocities and with coarse bed particle size, the fluid bed is dense with a uniform solids concentration, and it has a well-defined surface.

This system is called the *bubbling fluid bed* (BFB) because the air in excess of that required to fluidize the bed passes through the bed in the form of bubbles. This system has relatively low solids entrainment in the flue gas. With the *circulating fluid bed* (CFB) design, higher velocities and finer bed particle size are prevalent, and the fluid bed surface becomes diffuse as solids entrainment increases and there is no defined bed surface. The recycle of entrained material to the bed at high rates is required to maintain bed inventory. It is interesting that the BFB and CFB technologies are somewhat similar to stoker firing and pulverized coal firing with regard to fluidizing velocity, but the particle size of the bed is quite different. Stoker firing incorporates a fixed bed, has a comparable velocity, but has a much coarser particle size than that found in a BFB. For pulverizedcoal firing, the velocity is comparable with a CFB, but the particle size is much finer than that for a CFB.

Bubbling fluid bed (BFB) boiler

Of all the fluid bed technologies, the bubbling bed is the oldest. The primary difference between a BFB boiler and a CFB boiler design is that with a BFB the air velocity in the bed is maintained low enough that the material that comprises the bed (*e.g.*, fuel, ash, limestone, and sand), except for fines, is held in the bottom of the unit, and the solids do not circulate through the rest of the furnace enclosure. For new boilers, the BFB boilers are well suited to handle highmoisture waste fuels, such as sewage sludge, and also the various sludges that are produced in pulp and paper mills and in recycle paper plants. The features of design and the uniqueness of this technology, as well as the CFB. Although the boiler designs are different, the objectives of each are the same, and the designs are successful in achieving them.

Circulating fluid bed (CFB) boiler

The CFB boiler provides an alternative to stoker or pulverized coal firing. In general, it can produce steam up to 2 million lb/h at 2500 psig and 1000°F. It is generally selected for applications with high-sulfur fuels, such as coal, petroleum coke, sludge, and oil pitch, as well as for wood waste and for other biomass fuels such as vine clippings from large vineyards. It is also used for hard-to-burn fuels such as waste coal culm, which is a fine residue generally from the mining and production of anthracite coal. Because the CFB operates at a much lower combustion temperature than stoker or pulverized-coal firing, it generates approximately 50 percent less NO*x* as compared with stoker or pulverized coal firing. The use of CFB boilers is rapidly increasing in the world as a result of their ability to burn low-grade fuels while at the same time being able to meet the required emission criteria for nitrogen oxides (NO*x*), sulfur dioxide (SO2), carbon monoxide (CO), volatile organic compounds (VOC), and particulates. The CFB boiler produces steam economically for process purposes and for electric production. The advantages of a CFB boiler are reduced capital and operating costs that result primarily from the following:

1. It burns low-quality and less costly fuels.

2. It offers greater fuel flexibility as compared with coal-fired boilers and stoker-fired boilers.
3. It reduces the costs for fuel crushing because coarser fuel is used as compared with pulverized fuel. Fuel sizing is slightly less than that required for stoker firing.
4. It has lower capital costs and lower operating costs because additional pollution control equipment, such as SO2 scrubbers, is not required at certain site locations.,

COMBINED CYCLE AND COGENERATION SYSTEMS

In the 1970s and 1980s, the role of natural gas in the generation of electric power in the United States was far less than that of coal and oil. The reasons for this included

1. Low supply estimates of natural gas that projected it to last for less than 10 years
2. Natural gas distribution problems that threatened any reliable fuel delivery
3. Two OPEC (Organization of Petroleum Exporting Countries) oil embargoes that put pressure on the domestic natural gas supply
4. Concerns that natural gas prices would escalate rapidly and have an impact on any new exploration, recovery, and transmission

For these reasons, a Fuel Use Act was enacted in the late 1970s that prohibited the use of natural gas in new plants. This situation has changed dramatically because now the electric power industry is anticipating a continuing explosive growth in the use of natural gas. The reasons for this growth are

1. Continued deregulation of both natural gas and electric power
2. Environmental restrictions that limit the use of coal in many areas of the country
3. Continued perception of problems with the use of oil as a fuel for power plants because of greater dependence on foreign oil
4. Rapidly advancing gas turbine and combined cycle technology with higher efficiencies and lower emissions
5. Easier financing of power projects because of shorter schedules and more rapid return on investments

Perhaps the greatest reason for the growth is the current projection of natural gas supplies. Where before the natural gas supply was expected to last approximately 10 years, the current estimate is approximately 90 years based on the current production and use levels. Although this optimistic estimate is very favorable, it could promote a far greater usage, which could seriously deplete this critical resource in the future, far sooner than expected. Therefore, careful long-term plans must be incorporated for this energy source. This greater use of natural gas places additional demands on the natural gas pipeline industry.

Now, pipelines require regulatory approvals and also must accommodate any local opposition to a project. Most of the attention on the increased demand for natural gas has been focused on exploration and production of the fuel. Significantly less publicized but just as important is the need for handling this capacity with more capability for its delivery. This requires new pipelines to deliver the natural gas, new facilities to ship and receive liquefied natural gas (LNG), and additional underground aquifers or salt caverns for the storage of natural gas. North America has an extensive network of natural gas pipelines.

However, because of the projected demands, it is estimated that approximately 40,000 miles of new pipelines are required over the next decade. Many areas of the country are rejecting the addition of these pipelines in their area and thus causing additional cost and routing problems. Advancements in combustion technology have encouraged the application of natural gas to the generation of electric power. The gas turbine is the leader in combustion improvements. By using the most advanced metallurgy, thermal barrier coatings, and internal air cooling technology, the present-day gas turbines have higher outputs, higher reliability, lower heat rates, lower emissions, and lower costs. At present in the United States, nearly all new power plants that are fired by natural gas use gas turbines with combined cycles. Combined cycles (or cogeneration cycles) are a dual-cycle system. The initial cycle burns natural gas, and its combustion gases pass through a gas turbine that is connected to an electric generator. The secondary cycle is a steam cycle that uses the exhaust gases from the gas turbine for the generation of steam in a boiler.

The steam generated flows through a steam turbine that is connected to its electric generator. The interest in the combined cycle for power plants has resulted from the improved technology of gas turbines and the availability of natural gas. The steam cycle plays a secondary role in the system because its components are selected to match any advancement in technology such as the exhaust temperatures from gas turbines. The recovery of the heat energy from the gas turbine exhaust is the responsibility of the boiler, which for this combined cycle is called the *heat-recovery steam generator* (HRSG). As the exhaust temperatures from the more advanced gas turbines have increased, the design of the HRSG has become more complex.

The standard configuration of the HRSG is a vertically hung heat-transfer tube bundle with the exhaust gas flowing horizontally through the steam generator and with natural circulation for the water and steam. If required to meet emission regulations, selective catalytic reduction (SCR) elements for NOx control are placed between the appropriate tube bundles. This HRSG design also incorporates a duct burner (item 16). The duct burner is a system designed to increase high-pressure steam production from the HRSG. Its primary function is to compensate for the deficiencies of the gas turbine at high ambient temperature, especially during peak loads. The duct burner is seldom used during partial loads of the gas turbine and is not part of every HRSG design. The advantages of gas turbine combined cycle power plants are the following:

1. Modular construction results in the installation of large, highefficiency, base-loaded power plants in about 2 to 3 years.
2. Rapid, simple cycle startup of 5 to 10 minutes from no load to full load, which makes it ideal for peaking or emergency backup service.
3. High exhaust temperatures and gas flows enable the efficient use of heat-recovery steam generators for the cogeneration of steam and power.
4. Low NO_x and CO emissions.

ESSENTIALS OF STEAM POWER PLANT EQUIPMENT

A steam power plant must have following equipment:

- A furnace to burn the fuel.
- Steam generator or boiler containing water. Heat generated in the furnace is utilized to convert water into steam.
- Main power unit such as an engine or turbine to use the heat energy of steam and perform work.
- Piping system to convey steam and water.

In addition to the above equipment the plant requires various auxiliaries and accessories depending upon the availability of water, fuel and the service for which the plant is intended.

The flow sheet of a thermal power plant consists of the following four main circuits:

- Feed water and steam flow circuit.
- Coal and ash circuit.
- Air and gas circuit.
- Cooling water circuit.

A steam power plant using steam as working substance works basically on Rankine cycle.

Steam is generated in a boiler, expanded in the prime mover and condensed in the condenser and fed into the boiler again.

The different types of systems and components used in steam power plant are as follows:

- High pressure boiler
- Prime mover
- Condensers and cooling towers
- Coal handling system
- Ash and dust handling system
- Draught system
- Feed water purification plant
- Pumping system

- Air preheater, economizer, super heater, feed heaters.

The schematic arrangement of equipment of a steam power station. Coal received in coal storage yard of power station is transferred in the furnace by coal handling unit. Heat produced due to burning of coal is utilized in converting water contained in boiler drum into steam at suitable pressure and temperature. The steam generated is passed through the superheater. Superheated steam then flows through the turbine.

After doing work in the turbine the pressure of steam is reduced. Steam leaving the turbine passes through the condenser which is maintained the low pressure of steam at the exhaust of turbine. Steam pressure in the condenser depends upon flow rate and temperature of cooling water and on effectiveness of air removal equipment. Water circulating through the condenser may be taken from the various sources such as river, lake or sea. If sufficient quantity of water is not available the hot water coming out of the condenser may be cooled in cooling towers and circulated again through the condenser. Bled steam taken from the turbine at suitable extraction points is sent to low pressure and high pressure water heaters.

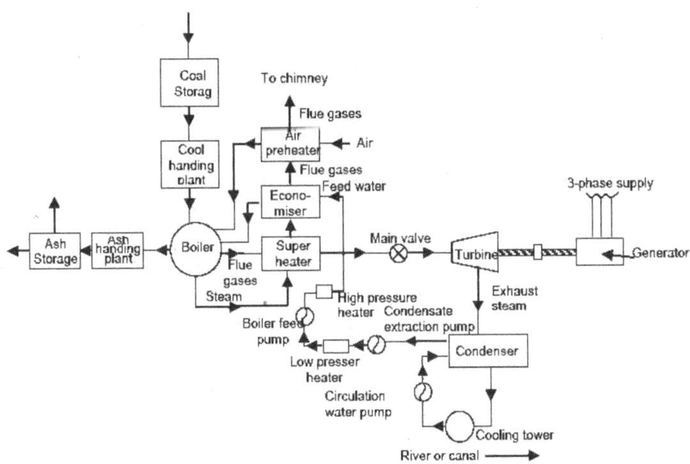

Fig. Steam power plant

Air taken from the atmosphere is first passed through the air pre-heater, where it is heated by flue gases. The hot air then passes through the furnace. The flue gases after passing over boiler and superheater tubes, flow through the dust collector and then through economiser, air pre-heater and finally they are exhausted to the atmosphere through the chimney.

Steam condensing system consists of the following:

- Condenser
- Cooling water
- Cooling tower
- Hot well

- Condenser cooling water pump
- Condensate air extraction pump
- Air extraction pump
- Boiler feed pump
- Make up water pump.

CLASSIFICATION

Boiler is an apparatus to produce steam. Thermal energy released by combustion of fuel is transferred to water, which vaporizes and gets converted into steam at the desired temperature and pressure.

The steam produced is used for:

- Producing mechanical work by expanding it in steam engine or steam turbine.
- Heating the residential and industrial buildings.
- Performing certain processes in the sugar mills, chemical and textile industries.

Boiler is a closed vessel in which water is converted into steam by the application of heat. Usually boilers are coal or oil fired. A boiler should fulfill the following requirements:

- Safety: The boiler should be safe under operating conditions.
- Accessibility: The various parts of the boiler should be accessible for repair and maintenance.
- Capacity: The boiler should be capable of supplying steam according to the requirements.
- Efficiency: To permit efficient operation, the boiler should be able to absorb a maximum amount of heat produced due to burning of fuel in the furnace.
- It should be simple in construction and its maintenance cost should be low.
- Its initial cost should be low.
- The boiler should have no joints exposed to flames.
- The boiler should be capable of quick starting and loading.

TYPES OF BOILERS

The boilers can be classified according to the following criteria. According to flow of water and hot gases:

- Water tube
- Fire tube

In water tube boilers, water circulates through the tubes and hot products of combustion flow over these tubes. In fire tube boiler the hot products of combustion pass through the tubes, which are surrounded, by water. Fire tube boilers have low initial cost, and are more compacts. But they are more likely to explosion, water volume is large and due to poor circulation they cannot meet quickly the change in steam demand. For the same output the outer shell of fire tube boilers is much larger than the shell of water-tube boiler. Water tube boilers require less weight of metal for a given size, are less liable to explosion, produce higher pressure, are accessible and can respond quickly to change in steam demand. Tubes and drums of water-tube boilers are smaller than that of fire-tube boilers and due to smaller size of drum higher pressure can be used easily. Water-tube boilers require lesser floor space. The efficiency of water-tube boilers is more. Water tube boilers are classified as follows:

Horizontal Straight Tube Boilers

- Longitudinal drum
- Cross-drum.

Bent Tube Boilers

- Two drum
- Three drum
- Low head three drum
- Four drum.

Cyclone Fired Boilers

Various advantages of water tube boilers are as follows:

- High pressure can be obtained.
- Heating surface is large. Therefore steam can be generated easily.
- Large heating surface can be obtained by use of large number of tubes.
- Because of high movement of water in the tubes the rate of heat transfer becomes large resulting into a greater efficiency.

Fire tube boilers are classified as follows:

- External Furnace:
- Horizontal return tubular
- Short fire box
- Compact.
- Internal Furnace
  - Horizontal Tubular
- Short firebox
- Locomotive
- Compact
- Scotch.
- Vertical Tubular
- Straight vertical shell, vertical tube
- Cochran (vertical shell) horizontal tube.

Various advantages of fire tube boilers are as follows:

- Low cost
- Fluctuations of steam demand can be met easily
- It is compact in size.

According to position of furnace:

- Internally fired
- Externally fired

In internally fired boilers the grate combustion chamber are enclosed within the boiler shell whereas in case of extremely fired boilers and furnace and grate are separated from the boiler shell.

According to the position of principle axis:

- Vertical
- Horizontal
- Inclined.

According to application:

- Stationary
- Mobile, (Marine, Locomotive).

According to the circulating water:

- Natural circulation
- Forced circulation.

According to steam pressure:

- Low pressure
- Medium pressure
- Higher pressure.

MAJOR COMPONENTS AND THEIR FUNCTIONS

Economizer

The economizer is a feed water heater, deriving heat from the flue gases. The justifiable cost of the economizer depends on the total gain in efficiency. In turn this depends on the flue gas temperature leaving the boiler and the feed water inlet temperature.

Air Pre-heater

The flue gases coming out of the economizer is used to preheat the air before supplying it to the combustion chamber. An increase in air temperature of 20 degrees can be achieved by this method. The pre heated air is used for combustion and also to dry the crushed coal before pulverizing.

Soot Blowers

The fuel used in thermal power plants causes soot and this is deposited on the boiler tubes, economizer tubes, air pre heaters, etc. This drastically reduces the amount of heat transfer of the heat exchangers. Soot blowers control the formation of soot and reduce its corrosive effects. The types of soot blowers are fixed type, which may be further classified into lane type and mass type depending upon the type of spray and nozzle used. The other type of soot blower is the retractable soot blower. The advantages are that they are placed far away from the high temperature zone, they concentrate the cleaning through a single large nozzle rather than many small nozzles and there is no concern of nozzle arrangement with respect to the boiler tubes.

Condenser

The use of a condenser in a power plant is to improve the efficiency of the power plant by decreasing the exhaust pressure of the steam below atmosphere. Another advantage of the condenser is that the steam condensed may be recovered to provide a source of good pure feed water to the boiler and reduce the water softening capacity to a considerable extent. A condenser is one of the essential components of a power plant.

Cooling Tower

The importance of the cooling tower is felt when the cooling water from the condenser has to be cooled. The cooling water after condensing the steam becomes hot and it has to be cooled as it belongs to a closed system. The Cooling towers do the job of decreasing the temperature of the cooling water after condensing the steam in the condenser. The type of cooling tower used in the Columbia Power Plant was an Inline Induced Draft Cross Flow Tower. This tower provides a horizontal air flow as the water falls down the tower in the form of small droplets. The fan centered at the top of units draws air through two cells that are paired to a suction chamber partitioned beneath the fan. The outstanding feature of this tower is lower air static pressure loss as there is less resistance to air flow. The evaporation and effective cooling of air is greater when the air outside is warmer and dryer than when it is cold and already saturated.

Superheater

The superheater consists of a superheater header and superheater elements. Steam from the main steam pipe arrives at the saturated steam chamber of the superheater header and is fed into the superheater elements. Superheated steam arrives back at the superheated steam chamber of the superheater header and is fed into the steam pipe to the cylinders. Superheated steam is more expansive.

Reheater

The reheater functions similar to the superheater in that it serves to elevate the steam temperature. Primary steam is supplied to the high pressure turbine. After passing through the high pressure turbine, the steam is returned to the steam generator for reheating (in a reheater) after which it is sent to the low pressure turbine. A second reheat cycle may also be provided.

10

Turbine Types and Applications

As described previously, a steam turbine takes the thermal energy of the steam, which is provided by a boiler, and converts it into useful mechanical work by means of the steam expanding as it flows through the turbine. Steam is introduced into the turbine through small stationary nozzles, where the steam expands and reaches a high velocity. This process converts the thermal energy in the steam to kinetic energy as it passes through the nozzle openings and moves the turbine blades, which are attached to the rotor. Steam turbines are designed for a variety of applications and in a variety of sizes that range from 1-hp (0.75-kW) units, which are used as drives for process equipment such as pumps and compressors, to large 1300-MW capacity units, which are found in large electric power plants. *Heat rate* is a term that is frequently used in the power industry, and it defines power plant efficiency. The net plant heat rate is the total fuel heat input in Btu/h divided by the net power output leaving the power plant in kilowatts. The net output is obtained by subtracting all auxiliary electric power needs of the plant from the gross output of the generator. The following equation expresses this relationship:

$$\text{Net plant heat rate (NPHR)} = \frac{\text{total fuel heat input (Btu/h)}}{\text{net electrical generation (kW)}}$$

The total energy efficiency of a plant can be expressed as follows:

$$\text{Efficiency, \%} = \frac{3413\ \text{Btu/kWh}}{\text{NPHR}} \times 100\%$$

The NPHR usually varies with the plant load and is expressed in Btu/kWh.

STEAM SUPPLY AND EXHAUST CONDITIONS.

When classifying steam turbines by their steam supply and exhaust conditions, they are categorized as condensing, Non-condensing or backpressure, reheat-condensing, and extraction and induction.

Condensing turbine

This type of steam turbine is used primarily as a drive for an electric generator in a power plant. These units exhaust steam at less than atmospheric pressure to a condenser. This extraction point withdraws steam that is used to heat feedwater in the feedwater heaters, or it can be used for some plant process.

Non-condensing or backpressure turbine

This type of turbine is used primarily in process plants, where the exhaust steam pressure is controlled by a regulating station that maintains the process steam at the required pressure. The unit shown can be designed for initial steam conditions of up to 1450 psi and 930°F, and it can produce outputs between 2 and 28 MW.

Reheat-condensing turbine

This type of turbine is used primarily in electricity-producing power plants. In these units, the main steam exhausts from the high-pressure section of the turbine and is returned to the boiler, where it is reheated with the associated increase in steam temperature. The steam is now at a lower pressure but often at the same superheat temperature as the initial steam conditions, and it is returned to the intermediate- and/or low-pressure sections of the turbine for further expansion.

Extraction and induction turbine

This type of turbine is also found primarily in process plants. On extraction turbines, steam is taken from the turbine at various extraction points and is used as process steam. In induction turbines, low-pressure steam is introduced into the unit at an intermediate stage to produce additional power. This extraction steam can be used for feedwater heating or for some process. This particular unit can be designed to handle steam with initial steam conditions as high as 2000 psi and 1000°F and provide electric outputs from 2 to 75 MW. For industrial-sized turbines as described above, the design must match the particular requirements of the plant. Some turbines are required to meet electric requirements and therefore are considered to be similar to large turbines for utility use in that they are condensing turbines. These units can be designed for electric outputs of up to approximately 150 MW and are housed in a single casing. Other industrial turbine designs furnish a specific quantity of exhaust steam or extraction steam for some industrial process or for heating. These turbines are called *backpressure turbines* and *extraction-condensing turbines.* It is this type of turbine that is used in combined cycle systems.

Casing or shaft arrangement

Steam turbines are also classified by their casing or shaft arrangement as being single, tandem-compound, or cross-compound and are described as follows:

1. *Single casing.* This is the basic arrangement for smaller units, where a single casing and shaft are used.
2. *Tandem-compound casing.* This arrangement has two or more casings on one shaft that drives a generator.
3. *Cross-compound casing.* This arrangement has two or more shafts that are not in line, with each shaft driving a generator. These units are found in large electric utility power plants.

TURBINE DESIGN AND CONSTRUCTION

A simple turbine consists of a shaft on which is mounted one or more disks. On the circumference of the disks are located blades or buckets to receive the steam and convert it into useful work. The rims of the disks have dovetail channels for receiving the blades. The ends of the blades are made to fit these dovetail channels. A turbine requires bearings for support, a suitable housing or casing to enclose the rotor, a system of lubrication, and a device known as a *governor* to maintain control over the speed. The efficiency and reliable performance of the turbine depend to a great extent on the design and construction of the blading.

The blades therefore must be made to withstand the action of the steam and the centrifugal force caused by the high speed at which the turbine must operate. In designing turbine blading, a compromise must be made between strength and economy. The stationary and moving elements have very little clearance, and any vibration of the moving element will cause them to rub. For this reason, extreme care must be exercised to design the rotors so that they will not vibrate. The length and size of the blades must be increased as the steam pressure drops and the steam increases in volume as it flows through the turbine. Large condensing turbines have large rotors and long buckets in their last stages.

This reduces the velocity of the steam leaving the turbine and as a result improves the efficiency of the turbine. Turbine blades are forgings made of steel or alloys, depending on the conditions under which they are to operate. The use of superheated steam requires blades that are specially designed to prevent warping and deterioration. Steam in the last stages of a turbine becomes very wet (approximately 10 to 15 percent moisture), and this moisture erodes the turbine blades. Special materials are used to lengthen their life. The blades are assembled on the disk, and a *shroud ring* is placed around their outer ends.

The tips of the blades pass through holes in the shroud ring. The ends are then welded so that they are held securely by the ring. When the blades are very long, extra lacing is sometimes used to tie them together to provide the necessary support. Rotors for small turbines consist of a machined-steel disk shrunk and keyed onto a heavy steel shaft. The shaft is rust protected at the gland zones by a sprayed coating of stainless steel. The rotor is statically and dynamically balanced to ensure smooth operation throughout its operating range. The small steam turbine is used as a mechanical drive. It contains a governor valve, a strainer, and an operating hand valve that is used for manual adjustment to obtain maximum efficiency. Regardless of whether or not a hand valve is provided, the steam is made to pass through the governor-controlled admission valve contained in the steam chest. These are only typical illustrations of a small turbine design, since there are numerous designs with different features that vary between manufacturers. The governor valve located in the right end of the steam chest is a double-seated balanced valve operating in a renewable cage. The valve and cage are made of Non-corrosive material, and the valve stem is made of stainless steel. Packing is provided where the valve stem passes through the chest cover. Speed regulation is maintained by means of the governor.

This governor is of the centrifugal-weight type and is connected to the admission valve by the arm. Movement of the governor weights is opposed by the compression spring to transmit its motion from the rotating spindle to a stationary sleeve through a self-aligning ball-thrust bearing. If the speed of the turbine exceeds a safe limit (10 percent above normal speed), an overspeed trip device is provided. This trip is eccentrically mounted and restrained by a spring to trip simultaneously the governor valve and the butterfly valve. Hand reset is provided when the speed has been reduced to normal. In normal operation, the butterfly valve is held open by the trip linkage against the force of the coil spring shown. In some cases the butterfly trip valve is omitted, and overspeeding closes the governor valve.

Located in the casing are the steam-admission nozzles, which are cut into a solid block of bronze or alloy steel, depending on steam conditions. Nozzles are so proportioned as to be contributory to efficient operation and are made of corrosion- and erosion-resistant materials. This nozzle block is bolted to the steam chest, which in turn is bolted to the base of the turbine casing. The entire assembly of nozzles for one stage is called a *diaphragm.* The casing assembly with the stationary blading or nozzles is referred to as the *turbine cylinder.* The cylinder of an impulse turbine is frequently referred to as the *wheel casing.* The turbine blades are made from rolled and drawn sections of stainless steel.

The blading, shrouding, and rotor rim are contoured to approximate the steam-expansion characteristics. The rotating blades are secured in dovetail grooves cut into the rotor disk and are regularly spaced by packing pieces. Around the outer rim of the blades or nozzles is provided a shroud ring to stiffen the blades against vibration and confine the steam to the blade path. Each group of blades is tied together. Round tenons at the blade ends are machine spun to secure shroud bands to the blades. Bearings support the rotor. They are horizontally split and are often lined with high-grade tin babbitt. Babbitt is a soft, silvery antifriction alloy composed of tin with small amounts of copper and antimony. Access to the bearings is possible without raising the cylinder cover or rotor. The governor-end bearing is babbitt-faced at both ends and acts as a combined thrust and journal bearing. Rings suspended in oil roll on the shaft to provide lubrication to the bearing. Water jackets are provided for cooling the oil. To prevent leakage of steam where the rotor shaft extends through the turbine casing, carbon-ring-type glands are provided. These consist of segmental rings held around the shaft by springs. Between the two outer rings of each gland is a leak-off connection that prevents steam that may pass the inner rings from leaking. The casing proper is bolted together at the horizontal joint. Flanges are frequently finish-ground, making possible a steamtight joint without the use of a gasket. If lubrication requirements exceed the capacity of a single pump, a separate gear-type pump is furnished to supply lubricating oil.

The direct-acting oil governor with hand-speed changer for 3:1 maximum speed adjustment is used for variable-speed applications, such as driving fans, blowers, etc. For the mechanically driven turbine, the speed is frequently automatically controlled by the draft, air pressure, feedwater pressure, and other means. *Note:* Design standards for materials and governor performance are covered in publications of the National Electrical Manufacturers Association (NEMA). On large turbines there are normally high- and low-pressure sections, with the steam chest integral with the high-pressure section. Cylinders are split along the horizontal plane. Blades may be assembled in separate blade rings or directly in the cylinder, depending on the turbine size and pressure and on temperature conditions. In order to minimize misalignment and distortion, turbines are so designed as to permit expansion and contraction in response to temperature changes. Larger turbines have their rotors formed from a singlepiece forging, including both the journals and the coupling flange.

Thrust-bearing collar and oil impeller may be carried on a stub shaft bolted to the end of the rotor. Forgings of this type are carefully heat treated and must conform to specifications. Rotors are machined, and after the blades are in place, they are dynamically balanced and tested. As with the small turbine, the efficiency and life of larger turbines are influenced chiefly by the form or shape of the blades, the manner in which they are fastened in place, and the materials from which they are made. *Rotating* blades are secured by a T-root fastening with lugs machined on the shank, straddling the blade groove. Blades are held against a shoulder in the groove by half-round sections caulked in place at the bottom. *Stationary* blades are anchored in straight-sided grooves by a series of short keys that fit into auxiliary grooves cut in the blade shank and in the side of the main groove. If high-pressure and high-temperature steam is used, steam leakage across impulse stages is controlled by thin sealing strips that can be set with close running clearance.

On reaction turbines of the larger sizes where highpressure and high-temperature steam is used, *shrouded* blades are used with radial seal strips to control leakage between stages. Seal strips are made very thin, permitting close running clearances. Two types of nozzles are used: round nozzles with holes drilled and reamed in a solid block of steel and curved-vane nozzles. The interstage diaphragms are located in grooves that are accurately spaced and machined into the casing. The upper halves are attached to, and lift with, the casing cover. Labyrinth seals minimize steam leaks along the shaft where it passes through the diaphragm. The seal rings are spring backed, and they are made of a material that permits close running clearances with complete safety. Turbine blades are at times designed in different configurations in an attempt by the designer to optimize performance and minimize maintenance. This blade design has reduced losses and has optimized the flow conditions for the blades. The advanced design shown has a forward-curved and twisted hollow blade.

This design provides optimal flow distribution. The hollow blade provides suction slots for moisture removal. By research and operating experience, each designer of turbines attempts to improve the overall performance and efficiency of the turbine. As a result, different designs have evolved. Main bearings consist of a cast-steel shell, split horizontally and lined with high-grade tin-base babbitt. They are supported in the bearing housing by steel blocks. Between the supporting blocks and the bearing shell are steel liners, where the bearings can be moved both vertically and horizontally to align the rotor accurately within the cylinder. The conventional turbine bearing consists of a cylindrical shell divided into halves so that it can be assembled on the shaft. The outside of the shell has a spherical section at the middle.

This fits into a similarly shaped seat in the bearing support pedestal and is an aid in properly aligning the bearing. In most cases the top half is grooved on the trailing side of the bearing in such a manner that there is a tendency to draw or aspirate oil into the bearing. The bottom half of a split bearing should be relieved on each side of the joint by scraping it clear of the shaft. The clearance of the bearing can be measured by placing a soft lead wire on the shaft and bolting the top half down tight. The flattened wire can be removed and the thickness measured by a micrometer. The clearance in the bearing should be 0.002 in per inch of shaft diameter. Each turbine manufacturer provides operating instructions and recommendations to ensure that clearances are established properly.

These should be followed. Oil is supplied through a hole drilled in the bearing housing that matches a similar hole in the lower supporting block. For large bearings, orifice outlets are provided for positive control. The pressure drop through the rotating blades of the reaction turbine produces a force that tends to push the turbine rotor Towards the low-pressure end. In some turbines this force is balanced by what is known as the *double-flow principle.* In this case, steam enters the middle of the turbine and flows in both directions. This produces two forces that balance each other.

THRUST BEARING

The thrust bearing consists of a collar rigidly attached to the turbine shaft rotating between two babbitt-lined shoes. The clearance between the collar and the shoes is small. The piston is attached to the spindle, and steam pressure is exerted on one side and atmospheric pressure is exerted on the other side. The difference in pressure produces a force that balances the thrust exerted on the rotating blades. If the shaft starts to move in either direction, the collar comes into contact with the shoes, and the shaft is held in proper position. Larger thrust bearings have several collars on the shaft and a corresponding number of stationary shoes. The *Kingsbury thrust bearing* is used when a large thrust load must be carried to maintain the proper axial position in the turbine cylinder.

The thrust collar is the same as that used in the common type of thrust bearing. The thrust shoes are made up of segments that are individually pivoted. With this arrangement, the pressure is distributed equally not only between the different segments but also on the individual segments. The openings between the segments permit the oil to enter the bearing surfaces. Almost 10 times as much pressure per square inch can be carried on the Kingsbury-type bearing as on the ordinary thrust bearing. Axial position of the bearing and turbine rotor may be adjusted by liners, located at the retainer rings, on each end of the bearing. The bearing is lubricated by circulating oil to all its moving parts.

The impulse turbine does not require as large a thrust bearing as the reaction turbine because there is little or no pressure drop through the rotating blades. However, the thrust bearing must be used to ensure proper clearance between the stationary and rotating elements. Reaction turbines that do not have some method of balancing the force caused by the drop in pressure in the rotating blades must be equipped with large thrust bearings. Turbine bearings are subjected to very severe service and require careful attention on the part of the operator. Most turbines operate at high speed (3600 rpm) and are subjected to the heat generated in the bearing itself as well as that received from the high-temperature steam. These conditions make necessary some method of cooling. In some cases the bearings are cooled by water jacketing; in others the oil is circulated through a cooler.

PACKING

The shaft at the high-pressure end of the turbine must be packed to prevent leakage of steam from the turbine. The one at the low-pressure end of a condensing turbine must be packed to prevent the leakage of air into the condenser. *Labyrinth* packing is used widely in steam turbine practice. It gets its name from the fact that it is so constructed that steam in leaking must follow a winding path and change its direction many times. This device consists of a drum that turns with the shaft and is grooved on the outside. The drum turns inside a stationary cylinder that is grooved on the inside.

There are many different types of labyrinth packing, but the general principle involved is the same for all. Steam in leaking past the packing is subjected to a throttling action. This action produces a reduction in pressure with each groove that the steam passes. The amount of leakage past the packing depends on the clearance between the stationary and the rotating elements. The amount of clearance necessary depends on the type of equipment, steam temperatures, and general service conditions. The steam that leaks past the labyrinth packing is piped to some lowpressure system or to a low stage on the turbine. A water-packed gland consists of a centrifugal-pump runner attached to the turbine shaft.

The runner rotates in a chamber in the gland casing. In some designs, water is supplied to the chamber at a pressure of 3 to 8 psi and is thrown out against the sides by the runner, forming a seal. Water seals are used in connection with labyrinth packing to prevent the steam that passes the packing from leaking into the turbine room. Such a seal is also used on the low-pressure end of condensing turbines. In this case the leakage to the condenser is water instead of air. They are used singly or in combination, depending on the service required.

Each labyrinth consists of a multiplicity of seals to minimize steam leakage. The seal rings are spring backed and made of material that permits close running clearances with safety. The glands are usually supplied with condensate water for sealing to prevent contamination of the condensate water. Seal designs are continuously being improved to minimize steam leakage and thus improve turbine performance. The illustrated designs are typical of those found on operating turbines. Carbon packing is composed of rings of carbon held against the shaft by means of springs. Each ring fits into a separate groove in the gland casing. When adjustments are made while the turbine is cold, carbon packing should have from 0.001 to 0.002 in of clearance per inch of shaft diameter.

The width of the groove in the packing casing should exceed the axial thickness of the packing ring by about 0.005 in. Carbon packing is sometimes used to pack the diaphragms of impulse turbines. Steam seals are used in connection with carbon packing. This is essential when carbon packing is used on the low-pressure end of condensing turbines, because if there is a slight packing leak, steam instead of air will leak into the condenser. In operating a turbine equipped with carbon packing, a slight leak is desirable because a small amount of steam keeps the packing lubricated. Flexible metallic packing is used to pack small single-stage turbines operating at low backpressure. In most cases the pressure in the casing of these turbines is only slightly above atmospheric pressure. The application is the same as when this packing is used for other purposes, except that care must be exercised in adjusting. Due to the high speed at which the shaft operates, even a small amount of friction will cause overheating.

GOVERNORS

A close control of turbine speed is essential from the standpoint of safety and satisfactory service. The same theory of centrifugal force that applied to steam engine flywheels also applies to the rotating elements of turbines. If the turbine runs at a speed far above that for which it was designed, the blading will be thrown out of the rotor. When a turbine is thrown apart in this manner, the resulting damage may be as great as, or even greater than, that caused by a boiler explosion. Turbines operating electric generators producing ac current must operate at constant speed (3600 or 3000 rpm). Some electrical appliances are seriously affected by a slight change of frequency in the power supply.

The speed of the generator determines the frequency of the electric-current generator (60 Hz for 3600 rpm and 50 Hz for 3000 rpm).1 Even with generators producing dc current, a small change in speed will affect the voltage. There are two ways of changing the turbine supply of steam to meet the load demand. One consists of throttling the steam, by means of a valve, in such a manner that the pressure on the first-stage elements is changed with the load demand. The operating mechanism is driven directly by the main turbine shaft. Overload is taken care of by means of hand-operated valves that admit additional steam to the turbine. The other method uses several valves, governor-operated, that are opened separately to supply steam to secondary nozzles as the load increases. Economical partial-load operation is obtained by minimizing throttling losses.

This is accomplished by dividing the first-stage nozzles into several groups and providing a separate valve to control the flow of steam to each group. Valves are then opened and closed in sequence, and the number of nozzle groups in service is proportional to the load on the unit. Valve seats are of the diffuser type to minimize pressure drop, and these valve seats are renewable. The multivalve steam chest is cast integrally with the cylinder cover with a cored passage from each valve to a nozzle group. Single-seated valves are used, arranged in parallel within the steam chest and surrounded by steam at throttle pressure. The governor mechanism raises and lowers the valve-lift bar in a horizontal plane, opening the valves in sequence, with an unbalanced force tending to close the valves. In the *oil-relay* system, the governor operates a small valve that admits oil to and allows it to drain from a cylinder. This oil cylinder contains a piston that is connected, by means of a rod, to the steamvalve mechanism.

The governor admits oil either above or below the piston, depending on which way the load is changing. There is no connection between the governor and the governor-valve mechanism except by means of the oil cylinder. The movement of the governor is transmitted to the governor valves by the oil pressure. Oil is supplied to the governor system by the same pump that supplies oil to the turbine bearings. If this oil supply should fail, the governor valve would close and stop the turbine, thus preventing damage to the bearings. The hydraulically operated throttle valve is used to control the flow of steam when starting a turbine and in addition functions as an automatic stop valve in case of overspeed. It cannot be opened nor can the turbine be started until after normal operating pressures for the turbine oiling system have been established. If oil pressure falls to an unsafe point, the valve automatically closes, and the flow of steam to the turbine is interrupted. A strainer is located within the valve to protect both the valve and the turbine. The emergency overspeed governor is separate and independent of the speed governor.

It functions to protect the unit from excessive speed by disengaging a trip at a predetermined speed, permitting the throttle valve to close. This trip may be reset and the throttle valve reopened before the turbine speed returns to normal. Large turbines are arranged frequently for extraction of steam at various points in the turbine. This extracted steam is used for feedwater heating or other heating or process purposes, and thus a more economical cycle is obtained. One turbine design uses a grid-type extraction valve at normal pressure and temperature. This valve consists of a stationary port ring and a rotating grid.

The rotating grid turns against the stationary ring and opens the ports in sequence. Thus simple hydraulic interconnections are obtained between the several components of the control system, such as the accurate and positive control of speed or the load carried by the unit, and of the extraction steam pressures. Where electric power generation is the prime consideration, condensing turbines are used. Openings are provided in the turbines for the extraction of steam for heating the feedwater to the boilers. This illustration is typical of an operating unit; however, turbine designs are continuously upgraded and vary between manufacturers. The change in speed of a turbine from no load to full load divided by the full-load speed is known as the *speed regulation.*

A turbine with a full-load speed of 1764 rpm and a no-load speed of 1800 rpm would have a change of 36 rpm, or regulation of 2 percent. These types of turbines are used to drive mechanical equipment such as pumps, compressors, etc. For electricity production, the speed must be constant at either 3600 or 3000 rpm depending on 60- or 50-Hz applications. The governors of turbines operating electric generators are supplied with synchronizing springs. These are arranged to aid in moving the governor weights and lowering the turbine speed or to work against the governor weights and increase the turbine speed.

The tension on the synchronizing spring is varied by means of a small motor (synchronizing motor). By adjusting the tension on this spring, the operator can change the load on the generator. *Dashpots* are frequently used in connection with turbine governors. They are used to keep the governor from overtraveling because of the weight of the governor parts. When a governor adjusts the speed of a turbine and makes a change that an adjustment in the opposite direction is immediately necessary, the governor is said to "hunt" or overtravel. In other cases dashpots are used to prevent the governor from operating too quickly. Friction or lost motion in the valve mechanism or, in some cases, too heavy or improperly installed dashpots will cause a turbine to hunt.

Rapid changes in speed and the corresponding variation in load are referred to as *surges*. A governor mechanism must be adjusted as follows: The valve should be closed when the governor is in the closed position and opened the correct amount when it is in the open position. The speed adjustment must be such that the governor will be in the open position when the turbine is at rated speed, and the speed regulation (*i.e.*, the change in speed from no load to full load) must be adjusted. Turbine governors require very little attention on the part of the operator. They must, however, be kept oiled and the joints kept working freely. With the oil-relay system, oil leaks must be stopped as soon as possible. The overspeed trip must be checked at regular intervals. It is good practice to operate the overspeed every time that the turbine is placed in service.

LUBRICATION

Proper lubrication is of utmost importance in the operation of a steam turbine. High journal speed, heat conducted from the steam to the bearing, and possible water leakage into the oil are some of the conditions that make lubrication difficult. There are two methods of lubricating turbine bearings. One utilizes oil rings in supplying oil to the bearings; the other consists of a pressure system that circulates the oil to the bearings. The oil-ring system is used on small turbines. There is an oil well under each bearing. The oil level in the bearing is kept below the bottom of the shaft. The oil rings have a larger diameter than that of the shaft.

They hang over the shaft, and a section extends into the oil. When the shaft rotates, the rings also turn and the oil that adheres to them is carried to the bearing. The bearings are grooved in such a manner that the oil is properly distributed to the bearing surfaces. This system is designed to supply bearing oil at 25 psig and hydraulic oil at 200 psig, approximately. It operates as follows: During the startup of the turbine, the steam-driven oil pump provides hydraulic oil to the control system, a portion of the hydraulic oil being fed to the booster-pump controls. Part of this oil pressure is reduced as the oil passes through the oil cooler to the bearing header for journal lubrication. The remainder of the oil to the booster pump drives an oil turbine coupled to the booster pump itself. The function of the booster pump is to provide suction pressure to the shaft-driven oil pump. When the main turbine reaches near-rated speed, the shaft-driven pump supplies the hydraulic oil to the controls as well as hydraulic oil to the booster pump. At this point, the steam-driven oil pump can be put in standby service. The bearing-oil pump is used to provide bearing oil only when the turbine is shut down and when it is operating on turning gear. During this period, oil is required for the bearing journals. With the governor in operation, the governor controls the pilot valve to position the control-valve piston that operates to open or close the steam-admission valves, thus regulating the speed of the turbine. Oil drained from journals and operating mechanism is returned to the oil reservoir, repeating the cycle. A variation of the foregoing comprises replacing the steam-driven oil pump with an ac motor-driven oil pump and, for emergency or standby service, the use of a dc emergency oil pump operating in parallel with the ac pump. The dc emergency unit is used when the turbine is operating below rated speed or when ac power is lost.

MAJOR COMPONENTS OF A TURBINE

The turbine consists of a shaft, which has one or more disks to which are attached moving blades, and a casing in which the stationary blades and nozzles are mounted. The shaft is supported within the casing by means of bearings that carry the vertical and circumference loads and by axial thrust bearings that resist the axial movement caused by the flow of steam through the turbine. Seals are provided in the casing to prevent the steam from bypassing the stages of the turbine.

Blades

On the outer portion, or circumference, of each disk located on the shaft are blades where steam is directed and converted into work by rotation of the shaft. There are many blades in each turbine stage, and larger turbines have more stages. Blades generally are made from lowcarbon stainless steel; however, for high-temperature applications and where high moisture is expected, alloy steels are used to provide the strength and erosion resistance needed. Special coatings on the blades are often used where high erosion is anticipated. As the steam flows through the turbine, it expands and its volume increases. This increased volume is handled by having longer blades and thus a larger casing for each stage of the turbine. The turbine efficiency, as well as its reliable performance, depends on the design and construction of the blades. Blades not only must handle the steam velocity and temperature but also must be able to handle the centrifugal force caused by the high speed of the turbine. Any vibration in a turbine is significant because there is little clearance between the moving blades and the stationary portions on the casing. A vibration of the moving blades could cause contact with the stationary components, which would result in severe damage to the turbine. Vibration has to be monitored continuously and corrected immediately when required.

Rotor shaft and bearings

The rotor shaft is supported at each end by bearings. These are normally ball bearings on small turbines; however, on the larger turbines, a pressure-lubricated journal bearing is used. Because of the axial thrust along the shaft that results from the difference in steam pressure across the stages of the turbine, thrust bearings are used to maintain the clearances between the moving blades and the stationary portions in each stage of the turbine.

Casings and seals

Casings are steel castings whose purpose is to support the rotor bearings and to have internal surfaces that will efficiently assist in the flow of steam through the turbine. The casing also supports the stationary blades and nozzles for all stages. At the turbine inlet, steam enters through a stop valve and steam chest. In high-temperature turbines these components are separate from the main turbine structure.

In smaller units, the steam chest is usually mounted directly on the casing. At the outlet of the turbine, an exhaust hood guides the steam from the last stage to the condenser inlet. This design has the stop valves located in the steam chest, and when necessary, this can be removed without having to dismantle piping.

The control valves are suspended on a cross-bar that is moved by two stems. Therefore, only two stuffing boxes are required for passing the stems through the casing, although four or five control valves are provided for nozzle group control. In addition to being designed to support the weight of the stationary nozzles and blades, the casing also must resist the mechanical stresses that are caused by the reaction forces on these nozzles and blades as well as the thermal stresses that are caused by the steam temperature differentials that occur during operation in the various stages of the turbine. Since the shaft penetrates through the casing, seals are necessary to minimize the leakage of steam. In small, lowtemperature turbines, carbon packing ring seals are used. These seals are located directly on the shaft and are held in place by a spring assembly. In larger turbines, labyrinth seals are used to control steam leakage. In many turbine designs, a combination of the two types of seals are used at the ends of the shaft.

Pressure sections of turbines

Small turbines are housed in a single casing that admits high-pressure steam at one end, and low-pressure steam leaves at the back end of the turbine to the condenser or as steam for a process or heating. On large, high-pressure turbines, two or three separate casings are used, with the turbine having three sections: high pressure (HP), intermediate pressure (IP), and low pressure (LP). The steam from the boiler passes through the HP section of the turbine. The exhaust steam from this section is returned to the boiler's reheater, where it is reheated to generally the same superheated steam temperature as the HP steam but at a lower pressure.

The steam is then returned to the IP section of the turbine. In some turbine designs, the HP and IP sections of the turbine are combined into one casing, and the steam flows in opposite directions in order to equalize the axial forces in the turbine. The IP turbine is used with boilers that have a reheat cycle, and this generally is found in large utility power plants. The IP turbine can use single or double flow depending on the pressure and steam flow. The LP turbine receives steam from either the HP turbine in Non-reheat units or an IP turbine. Since the steam has expanded and its volume has increased significantly, the blades of the LP turbine are much longer than those of the HP and IP stages in order to handle the steam flow.

Steam flow control

The part load performance and responsiveness of a steam turbine are to a large degree dependent on the method used to control the steam flow to the first stage of the turbine. There are two admission methods used: partial arc and full arc.

- *Partial-arc admission* requires the adjusting of the active nozzle area, where the nozzles in the first stage are divided into groups and controlled separately with throttling valves. When the load is increased, the throttling valves for each group of nozzles are opened in sequence to the full-throttle position until the desired load is attained.
- *Full-arc admission* requires the steam pressure entering all the firststage nozzles to be adjusted by either of two ways:
  - By operating the boiler at constant pressure and throttling the steam flow, with all valves being opened together until the required load is obtained
  - By varying the boiler operating pressure with the throttle valves basically

The selected control system is based on the planned operation of the turbine, since both systems have disadvantages. A base-loaded turbine would have a different control system than would a turbine planned for variations in load. Some of these disadvantages for each system are as follows:

1. Full-arc admission that requires constant throttle pressure is a relatively simple system, but when the turbine operates at part loads, pump power is wasted.
2. When the steam flow is throttled for full-arc admission, this results in a turbine inlet steam temperature drop.
3. When a partial-arc admission system is used that has a constant throttle pressure, this reduces the energy loss caused by steam throttling, but it is generally less efficient at full load.

4. By varying the boiler pressure, a less responsive system results for partial arc admission.

Therefore, each system affects the part load efficiency differently. Differences also exist on the turbine inlet steam temperature and on the system responsiveness. As a result, systems are designed with combined features of each control strategy in order to optimize the controls for the expected operation of the turbine.

TURBINE CONTROL

Turbine controls have evolved over the years to accommodate the increasingly complex operating requirements as well as the necessity for monitoring all major operating characteristics. A feature of this system is a high degree of system reliability through redundancy and self-diagnostic capabilities. Failure in one area of the system will not cause failure of the entire system.

The modularity of the design allows the addition of functions in increments. The base-control system shown as level 1 provides speed and load control and includes the manual backup and calibration unit. Redundant *distributed-processing units* (DPU) are provided for overspeed protection and operator autocontroller functions. Steam from the boiler enters the high-pressure turbine via throttle and governor valves. From the high-pressure turbine exhaust end, the steam flows to the reheater section of the boiler and reenters the intermediate-pressure portion of the turbine through reheat stop and interceptor valves.

The spring-loaded valves are positioned by hydraulic actuators that receive their fluid from a high-pressure fluid supply system. In the event of an overspeed condition, loss of pressure in the overspeed trip header is transmitted in the autostop trip header by means of a diaphragm valve. The digital controller positions the throttle and governing valves by means of electrohydraulic servo loops. In the event of a partial loop drop, the interceptor valves are closed by energizing a solenoid valve on the appropriate valve actuators. The digital controller receives three feedbacks from the turbine: speed, generator megawatt output, and first-stage steam pressure, which is proportional to the load. The operator controls the turbine and receives information from the manual panel or the operator's console, such as a *cathode-ray tube* (CRT), as well as an alarm and message printer.

ERECTION OF THE STEAM TURBINE

Steam turbines of less than 1 MW are usually shipped completely assembled. These units are assembled on their baseplates or bedplates and may be placed in position by means of rollers or a crane. In making hitches with a crane, care must be exercised to prevent appreciable deflection. The foundation must be roughed and swept clean to receive the grout. The turbine is set in position on iron blocks and tapered wedges. These must be arranged to hold the baseplate from 3D4 to 11D2 in from the foundation to allow space for the grouting.

The correct level is obtained by placing a sensitive level across machined bosses on the base. The level should be checked at all points. Small turbine units are assembled complete on one bedplate, but in the case of medium-sized units the turbine is separate from the machine that it is to drive. When the machines are separate, they must be aligned. The shafts must be central and parallel. The central relation of the two shafts may be checked by placing a straightedge parallel to the shafts and across the rims of the coupling flanges. This check should be applied at four places about the flanges. The parallel relation of the two shafts can be checked by measuring the distance between the coupling-flange faces with a feeler gauge. If the two shafts are parallel, these measurements will be equal at all points on the coupling.

Patent couplings that will compensate for a certain amount of misalignment are used. Even if these couplings are used, however, it is advisable to have the alignment as close as possible. New alignment technologies are now used and include the application of laser beams. *Manufacturer's alignment instructions should be followed explicitly.* The turbine changes in temperature more than the generator does, and if the shafts are lined up cold, they must be set to allow for expansion. Even if they are properly aligned when cold, this does not necessarily mean that they will be in proper alignment when in operation. The foundation bolts can now be tightened up and the level rechecked. If the level is found to be correct, the machine is ready to be grouted.

A dam of boards must be built around the foundation to keep the grouting in place. A suitable place must be provided for admitting the grout. Some foundations have holes for this purpose. The grouting mixtures vary from a pure cement to 2 parts sand and 1 part cement. The foundation should be wet before the grouting is poured. The mixture should be thin enough to flow readily so that it will find its way under the base. It is customary to leave the wedges in place, but in some cases they are removed after 11D2 or 2 days. Commercial grouting is readily available with recommended curing periods. Many medium-sized and large turbines are shipped completely disassembled because of equipment size limitations.

Their erection should be supervised by a thoroughly trained manufacturer's representative. Turbines require careful adjustment, and each different type presents different problems. The bearings are aligned by means of a fine steel line that is stretched through what is to be the center of the shaft. Laser beams are also used on the large turbine to obtain proper alignment. The axial clearance between stationary and moving parts must be checked with a taper gauge. The packing clearance must be checked and adjusted carefully. The piping to a steam turbine must be large enough to handle the steam flow at full load without excessive pressure drop. Steam velocities of 6000 ft/min were at one time considered the upper limit, but large high-pressure piping is now designed for velocities of 8000 or even 15,000 ft/min.

The piping must be arranged so that it will not produce strains on the turbine. This precaution is especially important for medium-sized and large turbines. Expansion is taken care of by means of loops or bends. The piping near the throttle valve is supported by springs that give with the expansion produced by the change from hot to cold condition yet hold the pipe firm enough to prevent vibration. A valve is placed in the steam line near the main header. This makes it possible to close off the line to the turbine, allowing work on the throttle valve while there is pressure on the main header. A strainer is often placed in the line to the turbine just ahead of the throttle valve. This prevents solid particles from entering the turbine.

RELIEF VALVES AND RUPTURE DISKS FOR TURBINES

A Non-condensing turbine with the exhaust line connected to an exhaust header must have a shutoff valve located near the header. The exhaust line also must have an expansion joint to prevent strains on the turbine or exhaust header. There must be a relief valve between the turbine and the shutoff valve. This reduces the possibility of a turbine explosion if, by mistake, the turbine should be operated with the shutoff valve closed. When a turbine is operated condensing, an atmospheric relief valve must be placed in the exhaust line between the turbine and the condenser. If the vacuum should fail, this valve would open and the turbine would exhaust to the atmosphere. It is necessary to keep a water seal on the atmospheric relief valve to prevent the leakage of air into the condenser.

Atmospheric relief valves are costly, and *rupture disks* or *rupture diaphragms* are used on some designs. A rupture disk is a prebulged membrane made of various metals based on the service for which it is intended. It is located in the exhaust hood of the turbine to prevent excessive pressure buildup if the condenser loses its vacuum. T

he disk may be used in lieu of an atmospheric relief valve or installed ahead of the atmospheric relief valve provided (1) the valve has ample capacity, (2) the maximum pressure range of the disk designed to rupture does not exceed the maximum allowable pressure of the vessel, (3) the area is at least equal to the area of the relief valve, or (4) the disk unit has a specified bursting pressure at a specific temperature and is guaranteed to burst within 5 percent (plus or minus) of its specific bursting pressure. If the pressure is greater than 15 psi, the ASME code states that either a rupture disk or a safety valve or a combination of the two may be used or the disk can be used in parallel with the safety valve so that the safety valve maintains all normal overpressure relief protection. Then the rupture disk is set at a higher pressure (approximately 20 percent above the safety relief valve). Therefore, if excessive pressure occurs, a safety system is ensured in case the safety relief valve fails. In all cases, the turbine designer must meet the ASME code requirements.

The rupture disk is fail-safe and ensures a seal-tight system with no moving parts and therefore nothing to malfunction, stick, or corrode shut. Therefore, it is a true safety fuse. A rupture disk cannot reclose itself, as can a safety relief valve. The rupture-disk assembly installed beneath the safety valve includes a telltale assembly consisting of a pressure gauge and an excess flow valve; therefore, leakage is immediately apparent. Jackscrews lift the discharge piping or outlet flange and provide for quick and easy rupture-disk replacement. Automatic vacuum breakers are also used to prevent the turbine from overspeeding. When the overspeed trip closes the throttle, there is the possibility that enough steam will leak past to cause the unit to overspeed. The emergency governor operates the vacuum breaker, which eliminates any possibility that enough steam will be drawn through the turbine to cause overspeeding.

TURBINE PIPING

The piping and turbine must be supplied with drains at all places where water can accumulate. The steam piping must have a drain located between the main steam line and the throttle. This drain is used to remove moisture from the line before placing the turbine in service. When saturated steam is used, a steam separator is placed in this line to remove the condensation. The drain is then connected to the separator. These drains are supplied with traps that are left open when the turbine is in operation and bypassed when it is being started.

The turbine casing must be drained, since an accumulation of water while the turbine is out of service would cause the blades to rust, and this would result in the rotor becoming unbalanced. A drain must be connected to the exhaust line of a Non-condensing turbine between the turbine and the shutoff valve. Drains must be supplied to carry the water away from the gland seals. These lines must be of sufficient size to handle an excess of water, because it is advisable to have this water discharge into funnels so that the flow can be observed by the operator.

The lines must be piped to a place where there is no backpressure. Mechanical-drive turbines operated below 250 psig are frequently used to start up (automatically) emergency electric generators in case of a power failure. They are also used to automatically start up pumps, air compressors, etc. For automatic operation, the throttle valve opens to place the turbine in service. The exhaust line remains open, and casing and other low points are provided with drain traps. Bearing cooling water is supplied either to an oil cooler, in the case of large turbines, or to water-jacketed bearings, in the case of small turbines. The cooling water for bearings should be as free from scaleforming material as possible. The water entering the drain lines should be visible to the operators. The discharge lines should be large enough to handle the discharge. The grouting of a new turbine must be allowed to set until sufficiently hard before the turbine is tried out. The piping and drain should be inspected carefully to see that the valves are properly arranged. It is good practice to blow out the steam lines with compressed air or auxiliary steam to remove mill scale and any foreign deposits that may be in the line.

TURBINE STAGE DESIGN

The efficiency of a turbine is optimized as the steam expands and does work in a number of steps or stages as it flows through the turbine. These stages are identified from the manner by which the energy is removed from the steam. There are two types of turbine stages, impulse and reaction, and most turbines combine features of both.

Impulse turbine

This stage design is often compared with a water wheel because nozzles direct the steam that flows through high-velocity jets. These steam jets, which contain kinetic energy, flow against the moving turbine blades or buckets. This energy is converted into mechanical energy by rotating the shaft. In a pure impulse turbine, when the steam passes through the stationary blades, it incurs a pressure drop. There is no pressure drop in the steam as it passes through the rotating blades. Therefore, in an impulse turbine, all the change of pressure energy into kinetic energy occurs in the stationary blades, while the change of kinetic energy into mechanical energy takes place in the moving blades of the turbine.

Reaction turbine

This design uses the reaction force resulting from the steam accelerating through the nozzles. The nozzles are actually created by the blades. Each stage of the turbine consists of a stationary set of blades and a row of rotating blades on a shaft. Since there is a continuous drop of pressure throughout each stage, steam is admitted around the entire circumference of the blades and, therefore, the stationary blades extend around the entire circumference. Steam passes through a set of stationary blades that direct the steam against the rotating blades. As the steam passes through these rotating blades, there is a pressure drop from the entrance side to the exit side that increases the velocity of the steam and produces rotation by the reaction of the steam on the blades.

Biblliography

Arthur P. Boresi and Richard J. Schmidt.: *Advanced Mechanics of Materials*, Wiley Publications, Delhi, 2008.

Braja M. Das.: *Advanced Soil Mechanics*, Taylor and Francis, Delhi, 2010.

D Jolly.: *Advanced Quantum Mechanics*, Sarup Publications, Delhi, 2006.

Eschenauer.: *Applied Structural Mechanics*, Springer Publications, Chennai, 2010.

Fomin, V.M., Kiselev, S.P., Vorozhtsov.: *Fluid Mechanics*, Jaico Publications, Delhi, 2009.

G D Arora.: *Analytical Mechanics*, Sarup Publications, Delhi, 2007.

Hilary Brewster.: *Fluid Mechanics*, Oxford Book Company, Jaipur, 2009.

J.A. Roberson and C.T. Crowe.: *Engineering Fluid Mechanics*, Jaico Publications, Delhi, 2007.

Javier E. Hasbun.: *Classical Mechanics with MATLAB Applications*, Jones and Bartlett Learning, 2010.

K. Purohit, S.P. Harsha, and R.K. Purohit.: *Fluid Mechanics*, Scientific Publications, Chennai, 2012.

Kibble.: *Classical Mechanics,* Cambridge University Press, Delhi, 2009.

Leif N. Persen.: *A Pragmatic Approach to Turbulence*: *A Short Course In Fluid Mechanics*, PHI Learning, Delhi, 2011.

Malcolm D. Bolton.: *A Guide to Soil Mechanics*, Universities Press, 2003.

Manish Saxena.: *Advances in Mechanics*, Oxford Book Company, Jaipur, 2012

Morin.: *Introduction to Classical Mechanics: With Problems and Solutions*, Cambridge University Press, Delhi, 2011.

Mrinal Chakraborty.: *Encyclopaedia of Quantum Mechanics*, Anmol Publications, Delhi, 2010.

Otto T. Bruhns.: *Advanced Mechanics of Solids*, Springer Publications, Chennai, 2008.

Pooja Bhagwan.: *A Handbook of Quantum Mechanics and Spectroscopy*, International Scientific Publications, 2005.

R. S. Khurmi.: *A Textbook of Applied Mechanics*, S. Chand Publisher, Delhi, 2001.

R.J. Garde and A.G. Mirajgaoker.: *Engineering Fluid Mechanics*, Scitech Publications, Delhi, 2010.

R.K. Dhawan.: *Applied Mechanics Engineering Mechanics*, S. Chand Publisher, 2011.

R.K.Rajput.: *A Textbook of Fluid Mechanics*, S. Chand Publisher, Delhi, 2002.

Rajesh Tripathi.: *Biomechanics of Human Motion*, Khel Sahitya Kendra, Delhi, 2010.

S Subrahmaniyam.: *Fluid Mechanics*, Capital Publishing Company, Kolkata, 2007.

S. S. Bhavikatti.: *A Textbook of Engineering Mechanics*, New Age Publications, Chennai, 2010.

Shashi Kant Yadav.: *Engineering Mechanics*, Discovery Publications, Delhi, 2006.

Shashi Kant Yadav.: *Textbook of Classical Mechanics*, Discovery Publications, Delhi, 2011.

T McClurg Anderson.: *Biomechanics of Human Motion*, Sports Publications, Delhi, 2007.

V.V. Vasiliev.: *Advanced Mechanics of Composite Materials*, Elsevier Publications, Delhi, 2010.

W. P. Graebel.: *Advanced Fluid Mechanics*, Elsevier Publications, Delhi, 2009.

Zohdi.: *An Introduction to Computational Micromechanics*, Springer Publications, Chennai, 2009.

Index